STOCHASTIC PROCESSES FOR PHYSICISTS

Understanding Noisy Systems

Stochastic processes are an essential part of numerous branches of physics, as well as biology, chemistry, and finance. This textbook provides a solid understanding of stochastic processes and stochastic calculus in physics, without the need for measure theory.

In avoiding measure theory, this textbook gives readers the tools necessary to use stochastic methods in research with a minimum of mathematical background. Coverage of the more exotic Levy processes is included, as is a concise account of numerical methods for simulating stochastic systems driven by Gaussian noise. The book concludes with a non-technical introduction to the concepts and jargon of measure-theoretic probability theory.

With over 70 exercises, this textbook is an easily accessible introduction to stochastic processes and their applications, as well as methods for numerical simulation, for graduate students and researchers in physics.

KURT JACOBS is an Assistant Professor in the Physics Department at the University of Massachusetts, Boston. He is a leading expert in the theory of quantum feedback control and the measurement and control of quantum nano-electro-mechanical systems.

T0344499

STOCHASTIC PROCESSES FOR PHYSICISTS

Understanding Noisy Systems

KURT JACOBS

University of Massachusetts

CAMBRIDGE
UNIVERSITY PRESS

CAMBRIDGE UNIVERSITY PRESS
Cambridge, New York, Melbourne, Madrid, Cape Town,
Singapore, São Paulo, Delhi, Mexico City

Cambridge University Press
The Edinburgh Building, Cambridge CB2 8RU, UK

Published in the United States of America by Cambridge University Press, New York

www.cambridge.org
Information on this title: www.cambridge.org/9780521765428

First published 2010
Reprinted with corrections 2013

A catalogue record for this publication is available from the British Library

ISBN 978-0-521-76542-8 Hardback

To Salman Habib and Bala Sundaram,
for pointing the way.

Contents

Preface

This book is intended for a one-semester graduate course on stochastic methods. It is specifically targeted at students and researchers who wish to understand and apply stochastic methods to problems in the natural sciences, and to do so without learning the technical details of measure theory. For those who want to familiarize themselves with the concepts and jargon of the "modern" measure-theoretic formulation of probability theory, these are described in the final chapter. The purpose of this final chapter is to provide the interested reader with the jargon necessary to read research articles that use the modern formalism. This can be useful even if one does not require this formalism in one's own research.

This book contains more material than I cover in my current graduate class on the subject at UMass Boston. One can select from the text various optional paths depending on the purpose of the class. For a graduate class for physics students who will be using stochastic methods in their research work, whether in physics or interdisciplinary applications, I would suggest the following: Chapters 1, 2, 3 (with Section 3.8.5 optional), 4 (with Section 4.2 optional, as alternative methods are given in 7.7), 5 (with Section 5.2 optional), 7 (with Sections 7.8 and 7.9 optional), and 8 (with Section 8.9 optional). In the above outline I have left out Chapters 6, 9 and 10. Chapter 6 covers numerical methods for solving equations with Gaussian noise, and is the sort of thing that can be picked-up at a later point by a student if needed for research. Chapter 9 covers Levy stochastic processes, including exotic noise processes that generate probability densities with infinite variance. While this chapter is no more difficult than the preceding chapters, it is a more specialized subject in the sense that relatively few students are likely to need it in their research work. Chapter 10, as mentioned above, covers the concepts and jargon of the rigorous measure-theoretic formulation of probability theory.

A brief overview of this book is as follows: Chapters 1 (probability theory without measure theory) and 2 (ordinary differential equations) give background material that is essential for understanding the rest of course. Chapter 2 will be almost all revision for students with an undergraduate physics degree. Chapter 3 covers all the basics of Ito calculus and solving stochastic differential equations. Chapter 4 introduces some further concepts such as auto-correlation functions, power spectra and white noise. Chapter 5 contains two applications (Brownian motion and option pricing), as well as a discussion of the Stratonovich formulation of stochastic equations and its role in modeling multiplicative noise. Chapter 6 covers numerical methods for solving stochastic equations. Chapter 7 covers Fokker–Planck equations. This chapter also includes applications to reaction–diffusion systems, and pattern formation in these systems. Chapter 8 explains jump processes and how they are described using master equations. It also contains applications to population dynamics and neuron behavior. Chapter 9 covers Levy processes. These include noise processes that generate probability densities with infinite variance, such as the Cauchy distribution. Finally Chapter 10 introduces the concepts and jargon of the "modern" measure-theoretic description of probability theory.

While I have corrected many errors that found their way into the manuscript, it is unlikely that I eliminated them all. For the purposes of future editions I would certainly be grateful if you can let me know of any errors you find.

Acknowledgments

Thanks to . . .

Aric Hagberg for providing me with a beautiful plot of labyrinthine pattern formation in reaction–diffusion systems (see Chapter 7) and Joshua Combes and Jason Ralph for bringing to my attention Edwin Jaynes' discussion of mathematical style (see Chapter 10). I am also grateful to my students for being patient and helping me iron-out errors and omissions in the text.

I am very grateful to Aaron Dörr, Andy Chia, Anita Dąbrowska, Chuck Yeung, Chris Chudzicki, David Pearson, David Sussillo, Jon Armond, Pu Ke, and Tom Chou, as well as the students in my graduate class, for sending me the errors they found in the first and second printings.

1

A review of probability theory

In this book we will study dynamical systems driven by noise. Noise is something that changes randomly with time, and quantities that do this are called *stochastic processes*. When a dynamical system is driven by a stochastic process, its motion too has a random component, and the variables that describe it are therefore also stochastic processes. To describe noisy systems requires combining differential equations with probability theory. We begin, therefore, by reviewing what we will need to know about probability.

1.1 Random variables and mutually exclusive events

Probability theory is used to describe a situation in which we do not know the precise value of a variable, but may have an idea of the relative likelihood that it will have one of a number of possible values. Let us call the unknown quantity X. This quantity is referred to as a *random variable*. If X is the value that we will get when we roll a six-sided die, then the possible values of X are $1, 2, \ldots, 6$. We describe the likelihood that X will have one of these values, say 3, by a number between 0 and 1, called the *probability*. If the probability that $X = 3$ is unity, then this means we will *always* get 3 when we roll the die. If this probability is zero, then we will never get the value 3. If the probability is $2/3$ that the die comes up 3, then it means that we expect to get the number 3 about two thirds of the time, if we roll the die many times.

The various values of X, and of any random variable, are an example of *mutually exclusive* events. That is, whenever we throw the die, X can have only one of the values between 1 and 6, no more and no less. Rather obviously, if the probability for X to be 3 is $1/8$, and for X to be 6 is $2/8$, then the probability for X to be *either* 3 *or* 6 is $1/8 + 2/8 = 3/8$. That is, the total probability that one of two or more mutually exclusive events occurs is the *sum* of the probabilities for each event. One usually states this by saying that "mutually exclusive probabilities sum". Thus, if

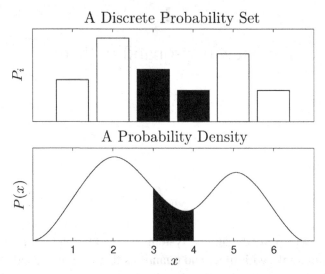

Figure 1.1. An illustration of summing the probabilities of mutually exclusive events, both for discrete and continuous random variables.

we want to know the probability for X to be in the range from 4 to 6, we sum all the probabilities for the values from 4 to 6. Figure 1.1 illustrates summing the probabilities from $X = 3$ to $X = 4$. Since X *always* takes a value between 1 and 6, the probability for it to take a value in this range must be unity. Thus, the sum of the probabilities for all the mutually exclusive possible values must always be unity. If the die is *fair*, then all the possible values are equally likely, and each is therefore equal to $1/6$.

Note: in mathematics texts it is customary to denote the unknown quantity using a capital letter, say X, and a variable that specifies one of the possible values that X may have as the equivalent lower-case letter, x. We will use this convention in this chapter, but in the following chapters we will use a lower-case letter for both the unknown quantity and the values it can take, since it causes no confusion.

In the above example, X is a *discrete random variable*, since it takes the discrete set of values $1, \ldots, 6$. If instead the value of X can be any real number, then we say that X is a *continuous* random variable. Once again we assign a number to each of these values to describe their relative likelihoods. This number is now a function of x (where x ranges over the values that X can take), called the *probability density*, and is usually denoted by $P_X(x)$ (or just $P(x)$). The probability for X to be in the range from $x = a$ to $x = b$ is now the area under $P(x)$ from $x = a$ to $x = b$. That is

$$\text{Prob}(a < X < b) = \int_a^b P(x)dx. \tag{1.1}$$

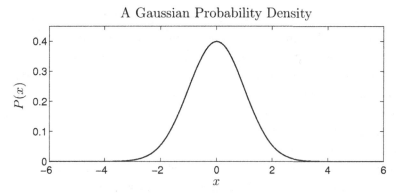

Figure 1.2. A Gaussian probability density with variance $V = 1$, and mean $\langle X \rangle = 0$.

This is illustrated in Figure 1.1. Thus the integral of $P(x)$ over the whole real line (from $-\infty$ to ∞) must be 1, since X must take one of these values:

$$\int_{-\infty}^{\infty} P(x)dx = 1. \tag{1.2}$$

The average of X, also known as the *mean*, or *expectation value*, of X is defined by

$$\langle X \rangle \equiv \int_{-\infty}^{\infty} P(x)x \, dx. \tag{1.3}$$

If $P(x)$ is symmetric about $x = 0$, then it is not difficult to see that the mean of X is zero, which is also the center of the density. If the density is symmetric about any other point, say $x = a$, then the mean is also a. This is clear if one considers a density that is symmetric about $x = 0$, and then shifts it along the x-axis so that it is symmetric about $x = a$: shifting the density shifts the mean by the same amount.

The *variance* of X is defined as

$$V_X \equiv \int_{-\infty}^{\infty} P(x)(x - \langle X \rangle)^2 \, dx = \int_{-\infty}^{\infty} P(x)x^2 \, dx - \langle X \rangle^2 = \langle X^2 \rangle - \langle X \rangle^2. \tag{1.4}$$

The *standard deviation* of X, denoted by σ_X and defined as the square root of the variance, is a measure of how broad the probability density for X is – that is, how much we can expect X to deviate from its mean value.

An important example of a probability density is the Gaussian, given by

$$P(x) = \frac{1}{\sqrt{2\pi\sigma^2}} e^{-\frac{(x-\mu)^2}{2\sigma^2}}. \tag{1.5}$$

The mean of this Gaussian probability density is $\langle X \rangle = \mu$ and the variance is $V(x) = \sigma^2$. A plot of this probability density in given in Figure 1.2.

1.2 Independence

Two random variables are referred to as being *independent* if neither of their probability densities depends on the value of the other variable. For example, if we rolled our six-sided die two times, and called the outcome of the first roll X, and the outcome of the second roll Y, then these two random variables would be independent. Further, we speak of the event $X = 3$ (when the first die roll comes up as 3) and the event $Y = 6$ as being independent. When two events are independent, the probability that both of them occur (that $X = 3$ *and* $Y = 6$) is the *product* of the probabilities that each occurs. One often states this by saying that "independent probabilities multiply". The reason for this is fairly clear if we consider first making the die roll to obtain X. Only if $X = 3$ do we then make the second roll, and only if that comes up 6 do we get the result $X = 3$ and $Y = 6$. If the first roll only comes up 3 one eighth of the time, and the second comes up 6 one sixth of the time, then we will only get both of them $1/8 \times 1/6 = 1/48$ of the time.

Once again this is just as true for independent random variables that take a continuum of values. In this case we speak of the "joint probability density", $P(x, y)$, that X is equal to x and Y is equal to y. This joint probability density is the product of the probability densities for each of the two independent random variables, and we write this as $P(x, y) = P_X(x) P_Y(y)$. The probability that X falls within the interval $[a, b]$ *and* Y falls in the interval $[c, d]$ is then

$$\text{Prob}(X \in [a, b] \text{ and } Y \in [c, d]) = \int_a^b \int_c^d P(x, y) dy dx$$

$$= \int_a^b \int_c^d P_X(x) P_Y(y) dy dx = \left(\int_a^b P_X(x) dx \right) \left(\int_c^d P_Y(y) dy \right)$$

$$= \text{Prob}(X \in [a, b]) \times \text{Prob}(Y \in [c, d]).$$

In general, if we have a joint probability density, $P(x_1, \ldots, x_N)$, for the N variables $X_1, \ldots, X_N$, then the expectation value of a function of the variables, $f(X_1, \ldots, X_N)$, is given by integrating the joint probability density over all the variables:

$$\langle f(X_1, \ldots, X_N) \rangle = \int_{-\infty}^{\infty} f(x_1, \ldots, x_N) P(x_1, \ldots, x_N) dx_1 \ldots dx_N. \quad (1.6)$$

It is also worth noting that when two variables are independent, then the expectation value of their product is simply the product of their individual expectation values. That is

$$\langle XY \rangle = \langle X \rangle \langle Y \rangle. \quad (1.7)$$

1.3 Dependent random variables

Random variables, X and Y, are said to be *dependent* if their joint probability density, $P(x, y)$, does not factor into the product of their respective probability densities.

To obtain the probability density for one of the variables alone (say X), we integrate the joint probability density over all values of the other variable (in this case Y). This is because, for each value of X, we want to know the total probability summed over all the mutually exclusive values that Y can take. In this context, the probability densities for the single variables are referred to as the *marginals* of the joint density.

If we know nothing about the value of Y, then our probability density for X is just the marginal

$$P_X(x) = \int_{-\infty}^{\infty} P(x, y)dy. \tag{1.8}$$

If X and Y are dependent, and we learn the value of Y, then in general this will change our probability density for X (and vice versa). The probability density for X *given* that we know that $Y = y$, is written $P(x|y)$, and is referred to as the *conditional* probability density for X *given* Y.

To see how to calculate this conditional probability, we note first that $P(x, y)$ with $y = a$ gives the *relative* probability for different values of x given that $Y = a$. To obtain the conditional probability density for X given that $Y = a$, all we have to do is divide $P(x, a)$ by its integral over all values of x. This ensures that the integral of the conditional probability is 1. Since this is true for any value of y, we have

$$P(x|y) = \frac{P(x, y)}{\int_{-\infty}^{\infty} P(x, y)dx}. \tag{1.9}$$

Note also that since

$$P_Y(y) = \int_{-\infty}^{\infty} P(x, y)dx, \tag{1.10}$$

if we substitute this into the equation for the conditional probability above (Eq. (1.9)) we have

$$P(x|y) = \frac{P(x, y)}{P_Y(y)}, \tag{1.11}$$

and further that $P(x, y) = P(x|y)P_Y(y)$.

As an example of a conditional probability density consider a joint probability density for X and Y, where the probability density for Y is a Gaussian with zero mean, and that for X is a Gaussian whose mean is given by the value of Y. In this case X and Y are not independent, and we have

$$P(x, y) = P(x|y)P(y) = \frac{e^{-(1/2)(x-y)^2}}{\sqrt{2\pi}} \times \frac{e^{-(1/2)y^2}}{\sqrt{2\pi}} = \frac{e^{-(1/2)(x-y)^2-(1/2)y^2}}{2\pi},$$

(1.12)

where we have chosen the variance of Y, and of X given Y to be unity. Generally, when two random variables are dependent, $\langle XY \rangle \neq \langle X \rangle \langle Y \rangle$.

1.4 Correlations and correlation coefficients

The expectation value of the product of two random variables is called the *correlation* of the two variables. The reason that we call this quantity a *correlation* is that, if the two random variables have zero mean and fixed variance, then the larger the value of the correlation, the more the variables tend to fluctuate *together* rather than independently; that is, if one is positive, then it is more likely that the other is positive. The value of the correlation therefore indicates how *correlated* the two variables are.

Of course, if we increase the variance of either of the two variables then the correlation will also increase. We can remove this dependence, and obtain a quantity that is a clearer indicator of the mutual dependence between the two variables by dividing the correlation by $\sqrt{V(X)V(Y)}$. This new quantity is called the correlation *coefficient* of X and Y, and is denoted by C_{XY}:

$$C_{XY} \equiv \frac{\langle XY \rangle}{\sqrt{V(X)V(Y)}}.$$

(1.13)

If the means of X and Y are not zero, then we can remove these when we calculate the correlation coefficient, so as to preserve its properties. Thus, in general, the correlation coefficient is defined as

$$C_{XY} \equiv \frac{\langle (X - \langle X \rangle)(Y - \langle Y \rangle) \rangle}{\sqrt{V(X)V(Y)}} = \frac{\langle XY \rangle - \langle X \rangle \langle Y \rangle}{\sqrt{V(X)V(Y)}}.$$

(1.14)

The quantity on the top line, $\langle XY \rangle - \langle X \rangle \langle Y \rangle$ is called the *covariance* of X and Y, and is zero if X and Y are independent. The correlation coefficient is therefore zero if X and Y are independent (completely uncorrelated), and is unity if $X = cY$, for some positive constant c (perfect correlation). If $X = -cY$, then the correlation coefficient is -1, and we say that the two variables are perfectly *anti-correlated*. The correlation coefficient provides a rough measure of the mutual dependence of two random variables, and one which is relatively easy to calculate.

1.5 Adding independent random variables together

When we have two independent continuous random variables, X and Y, with probability densities P_X and P_Y, it is often useful to be able to calculate the probability density of the random variable whose value is the sum of them: $Z = X + Y$. It turns out that the probability density for Z is given by

$$P_Z(z) = \int_{-\infty}^{\infty} P_X(z - s)P_Y(s)ds \equiv P_X * P_Y, \qquad (1.15)$$

which is called the *convolution* of P_X and P_Y [1]. Note that the convolution of two functions, denoted by "$*$", is another function. It is, in fact, quite easy to see directly why the above expression for $P_Z(z)$ is true. For Z to equal z, then if $Y = y$, X must be equal to $z - y$. The probability (density) for that to occur is $P_Y(y)P_X(z - y)$. To obtain the total probability (density) that $Z = z$, we need to sum this product over all possible values of Y, and this gives the expression for $P_Z(z)$ above.

It will be useful to know the mean and variance of a random variable that is the sum of two or more random variables. It turns out that if $X = X_1 + X_2$, then the mean of X is

$$\langle X \rangle = \langle X_1 \rangle + \langle X_2 \rangle, \qquad (1.16)$$

and if X_1 and X_2 are independent, then

$$V_X = V_{X_1} + V_{X_2}. \qquad (1.17)$$

That is, when we add independent random variables both the means and variances add together to give the mean and variance of the new random variable. It follows that this remains true when we add any number of independent random variables together, so that, for example, $\langle \sum_{n=1}^{N} X_n \rangle = \sum_{n=1}^{N} \langle X_n \rangle$.

If you have ever taken an undergraduate physics lab, then you will be familiar with the notion that averaging the results of a number of independent measurements produces a more accurate result. This is because the variances of the different measurement results add together. If all the measurements are made using the same method, we can assume the results of all the measurements have the same mean, μ, and variance, V. If we average the results, X_n, of N of these independent measurements, then the mean of the average is

$$\mu_{\text{av}} = \left\langle \sum_{n=1}^{N} \frac{X_n}{N} \right\rangle = \sum_{n=1}^{N} \frac{\mu}{N} = \mu. \qquad (1.18)$$

But because we are dividing each of the variables by N, the variance of each goes down by $1/N^2$. Because it is the variances that add together, the variance of the sum is

$$V_{\text{av}} = V\left[\sum_{n=1}^{N} \frac{X_n}{N}\right] = \sum_{n=1}^{N} \frac{V}{N^2} = \frac{V}{N}. \tag{1.19}$$

Thus the variance gets smaller as we add more results together. Of course, it is not the variance that quantifies the uncertainty in the final value, but the standard deviation. The standard deviation of each measurement result is $\sigma = \sqrt{V}$, and hence the standard deviation of the average is

$$\sigma_{\text{av}} = \sqrt{\frac{V}{N}} = \frac{\sigma}{\sqrt{N}}. \tag{1.20}$$

The accuracy of the average therefore increases as the square root of the number of measurements.

1.6 Transformations of a random variable

If we know the probability density for a random variable X, then it can be useful to know how to calculate the probability density for a random variable, Y, that is some function of X. This is referred to as a *transformation* of a random variable because we can think of the function as transforming X into a new variable Y. Let us begin with a particularly simple example, in which Y is a linear function of X. This means that $Y = aX + b$ for some constants a and b. In this case it is not that hard to see the answer directly. Since we have multiplied X by a, the probability density will be stretched by a factor of a. Then adding b will shift the density by b. The result is that the density for Y is $Q(Y) = P(y/a - b/a)/a$.

To calculate the probability density for $Y = aX + b$ in a more systematic way (which we can then use for much more general transformations of a random variable) we use the fact that the probability density for Y determines the average value of a function of Y, $f(Y)$, through the relation

$$\langle f(Y)\rangle = \int_{-\infty}^{\infty} P(y)f(y)dy. \tag{1.21}$$

Now, since we know that $Y = g(X) = aX + b$, we also know that

$$\langle f(Y)\rangle = \int_{-\infty}^{\infty} P(x)f(y)dx = \int_{-\infty}^{\infty} P(x)f(ax + b)dx. \tag{1.22}$$

Changing variables in the integral from x to y we have

$$\langle f(Y)\rangle = \int_{-\infty}^{\infty} P(x)f(ax + b)dx = \frac{1}{a}\int_{-\infty}^{\infty} P(y/a - b/a)f(y)dy. \tag{1.23}$$

Thus the density for Y is

$$Q(y) = \frac{1}{a}P(y/a - b/a). \tag{1.24}$$

In addition, it is simple to verify that $\langle Y \rangle = a\langle X \rangle + b$ and $V_Y = a^2 V_X$.

More generally, we can derive an expression for the probability density of Y when Y is an arbitrary function of a random variable. If $Y = g(X)$, then we determine the probability density for Y by changing variables in the same way as above. We begin by writing the expectation value of a function of Y, $f(Y)$, in terms of $P(x)$. This gives

$$\langle f(Y) \rangle = \int_{x=a}^{x=b} P(x)f(g(x))dx, \tag{1.25}$$

where a and b are, respectively, the lower and upper limits on the values that X can take. Now we transform this to an integral over the values of Y. Denoting the inverse of the function g as g^{-1}, so that $X = g^{-1}(Y)$, we have

$$\langle f(Y) \rangle = \int_{x=a}^{x=b} P(x)f(g(x))dx = \int_{y=g(a)}^{y=g(b)} P(g^{-1}(y))\left(\frac{dx}{dy}\right)f(y)dy$$

$$= \int_{y=g(a)}^{y=g(b)} \frac{P(g^{-1}(y))}{g'(x)}f(y)dy = \int_{y=g(a)}^{y=g(b)} \frac{P(g^{-1}(y))}{g'(g^{-1}(y))}f(y)dy. \tag{1.26}$$

We now identify the function that multiplies $f(y)$ inside the integral over y as the probability density for Y. But in doing so we have to be a little bit careful. If the lower limit for y, $g(a)$, is *greater* than the upper limit for y, then the probability density we get will be negative to compensate for this inversion of the integral limits. So the probability density is actually the absolute value of the function inside the integral. The probability density for y is therefore

$$Q(y) = \frac{P(g^{-1}(y))}{|g'(g^{-1}(y))|}. \tag{1.27}$$

One must realize also that this expression for $Q(y)$ only works for functions that map a single value of x to a single value of y (invertible functions), because in the change of variables in the integral we assumed that g was invertible. For non-invertible functions, for example $y = x^2$, one needs to do the transformation of the integral on a case-by-case basis to work out $Q(y)$.

1.7 The distribution function

The *probability distribution function*, which we will call $D(x)$, of a random variable X is defined as the probability that X is less than or equal to x. Thus

$$D(x) = \text{Prob}(X \leq x) = \int_{-\infty}^{x} P(z)\,dz. \tag{1.28}$$

In addition, the fundamental theorem of calculus tells us that

$$P(x) = \frac{dD(x)}{dx}. \tag{1.29}$$

1.8 The characteristic function

Another very useful definition is that of the *characteristic* function, $\chi(s)$. The characteristic function is defined as the *Fourier transform* of the probability density. Thus before we discuss the characteristic function, we need to explain what the Fourier transform is. The Fourier transform of a function $P(x)$ is another function given by

$$\chi(s) = \int_{-\infty}^{\infty} P(x)e^{isx}dx. \tag{1.30}$$

The Fourier transform has many useful properties. One of them is the fact that it has a simple inverse, allowing one to perform a transformation on $\chi(s)$ to get back $P(x)$. This inverse transform is

$$P(x) = \frac{1}{2\pi} \int_{-\infty}^{\infty} \chi(s)e^{-isx}ds. \tag{1.31}$$

Another very useful property is the following. If we have two functions $F(x)$ and $G(x)$, then the Fourier transform of their convolution is simply the *product* of their respective Fourier transforms! This can be very useful because a product is always easy to calculate, but a convolution is not. Because the density for the sum of two random variables in the convolution of their respect densities, we now have an alternate way to find the probability density of the sum of two random variables: we can either convolve their two densities, or we can calculate the characteristic functions for each, multiply these together, and then take the inverse Fourier transform.

Showing that the Fourier transform of the convolution of two densities is the product of their respective Fourier transforms is not difficult, but we do need to use the Dirac δ-function, denoted by $\delta(x)$. The Dirac δ-function is zero everywhere except at $t = 0$, where it is infinite. It is defined in such a way that it integrates to unity:

$$\int_{-\infty}^{\infty} \delta(x)dx = 1. \tag{1.32}$$

We get the δ-function if we take the limit in which a function with fixed area becomes increasingly sharply peaked about $x = 0$. If we shift the δ-function so that it is peaked at $x = c$, multiply it by another function, $f(x)$, and integrate over all space, this picks out the value of $f(x)$ at $x = c$:

$$\int_{-\infty}^{\infty} \delta(x - c)f(x)dx = f(c). \tag{1.33}$$

The δ-function can be rigorously defined using the theory of distributions (which was introduced for this purpose) or using measure theory. The δ-function is very useful when using Fourier transforms. The δ-function is the Fourier transform of the constant function $f = 1/(2\pi)$. That is

$$\frac{1}{2\pi} \int_{-\infty}^{\infty} e^{isx}dx = \delta(s). \tag{1.34}$$

With the above results we can now show that the Fourier transform of the convolution of two functions, $P(x)$ and $Q(x)$, is the product of their respective Fourier transforms, $\chi_P(s)$ and $\chi_Q(s)$. Denoting the convolution of $P(x)$ and $Q(x)$ as $R(x) = P * Q$, and using the definition of the Fourier transform, we have

$$\chi_R(s) = \int_{-\infty}^{\infty} R(x)e^{isx}dx = \int \left[\int P(y)Q(x - y)dy \right] e^{isx}dx$$

$$= \frac{1}{(2\pi)^2} \iiint \chi_P(s')e^{-is'y}\chi_Q(s'')e^{-is''(x-y)}e^{isx}ds'ds''dydx$$

$$= \frac{1}{(2\pi)^2} \iint \left[\int e^{ix(s-s'')}dx \int e^{iy(s''-s')}dy \right] \chi_P(s')\chi_Q(s'')ds'ds''$$

$$= \iint \delta(s - s'')\delta(s'' - s')\chi_P(s')\chi_Q(s'')ds'ds''$$

$$= \int \delta(s - s'')\chi_P(s'')\chi_Q(s'')ds'' = \chi_P(s)\chi_Q(s), \tag{1.35}$$

where all the integrals are from $-\infty$ to ∞.

One can also define the characteristic function for discrete random variables, and it has all the same properties. To do this, we again use the handy δ-function. Let us say that we have a discrete random variable X, that takes the values α_n, for $n = 1, 2, \ldots, N$. If the probabilities for the values α_n are P_n, then we can write a *probability density* for X using δ-functions at the locations α_n. This probability density for X is

$$P(x) = \sum_{n=1}^{N} p_n \delta(x - \alpha_n). \tag{1.36}$$

The characteristic function for X is then the Fourier transform of this probability density:

$$\chi_X(s) = \int_{-\infty}^{\infty} \left[\sum_{n=1}^{N} p_n \delta(x - \alpha_n) \right] e^{isx} dx = \sum_{n=1}^{N} \exp\{is\alpha_n\} p_n. \tag{1.37}$$

Note that regardless of whether the characteristic function is for a continuous or discrete random variable, one can always write it as the expectation value

$$\chi_X(s) = \left\langle e^{isX} \right\rangle, \tag{1.38}$$

where X is the random variable. For a probability density for a set of random variables, $X_1, \ldots, X_N$, the characteristic function is

$$\chi_{\mathbf{X}}(\mathbf{s}) = \left\langle e^{i\mathbf{s} \cdot \mathbf{X}} \right\rangle, \tag{1.39}$$

where $\mathbf{X} = (X_1, \ldots, X_N)$ is the vector of random variables, and $\mathbf{s} = (s_1, \ldots, s_N)$.

1.9 Moments and cumulants

For a random variable X, the expectation value of X^n, $\langle X^n \rangle$, is called the *n*th *moment*. The moments can be calculated from the derivatives of the characteristic function, evaluated at $s = 0$. We can see this by expanding the characteristic function as a Taylor series:

$$\chi(s) = \sum_{n=0}^{\infty} \frac{\chi^{(n)}(0) s^n}{n!}, \tag{1.40}$$

where $\chi^{(n)}(s)$ is the *n*th derivative of $\chi(s)$. But we also have

$$\chi(s) = \left\langle e^{isX} \right\rangle = \left\langle \sum_{n=0}^{\infty} \frac{(isX)^n}{n!} \right\rangle = \sum_{n=0}^{\infty} \frac{(i)^n \langle X^n \rangle s^n}{n!}. \tag{1.41}$$

Equating these two expressions for the characteristic function gives us

$$\langle X^n \rangle = \frac{\chi^{(n)}(0)}{(i)^n}. \tag{1.42}$$

The *n*th-order *cumulant* of X, κ_n, which is a polynomial in the first n moments, is also given by Eq. (1.42), but with χ replaced with $\log(\chi)$. The reason for this apparently rather odd definition is that it gives the cumulants a special property.

If we add two random variables together, then the *n*th cumulant of the result is the *sum* of the *n*th cumulants of the two random variables: when we add random variables, all the cumulants merely add together. The first cumulant is simply the mean, and the second cumulant is the variance. The next two cumulants are given by

$$\kappa_3 = \langle X^3 \rangle - 3\langle X^2 \rangle \langle X \rangle + 2\langle X \rangle^3, \tag{1.43}$$

$$\kappa_4 = \langle X^4 \rangle - 3\langle X^2 \rangle^2 - 4\langle X^3 \rangle \langle X \rangle + 12\langle X^2 \rangle \langle X \rangle^2 - 6\langle X \rangle^4. \tag{1.44}$$

Note that the Gaussian probability density is special, because all cumulants above second order vanish. This is because the characteristic function for the Gaussian is also a Gaussian. Taking the log cancels out the exponential in the Gaussian, and we are left with a quadratic in *s*, so that the Taylor series stops after $n = 2$.

For probability densities that contain more than one variable, say *x* and *y*, the moments are defined as $\langle X^n Y^m \rangle$. When $m = 0$, these are the moments for *X* alone. When $n = m = 1$ this is the correlation of *X* and *Y*, as defined in Section 1.2 above.

1.10 The multivariate Gaussian

It is possible to have a probability density for *N* variables, in which the marginal densities for each of the variables are all Gaussian, and where all the variables may be correlated. Defining a column vector of *N* random variables, $\mathbf{x} = (x_1, \ldots, x_N)^{\mathrm{T}}$, the general form of this *multivariate* Gaussian is

$$P(\mathbf{x}) = \frac{1}{\sqrt{(2\pi)^N \det[\Gamma]}} \exp\left[-\frac{1}{2}(\mathbf{x} - \boldsymbol{\mu})^{\mathrm{T}} \Gamma^{-1} (\mathbf{x} - \boldsymbol{\mu}) \right]. \tag{1.45}$$

Here $\boldsymbol{\mu}$ is the vector of the means of the random variables, and Γ is the matrix of the covariances of the variables,

$$\Gamma = \langle \mathbf{X}\mathbf{X}^{\mathrm{T}} \rangle - \langle \mathbf{X} \rangle \langle \mathbf{X}^{\mathrm{T}} \rangle = \langle \mathbf{X}\mathbf{X}^{\mathrm{T}} \rangle - \boldsymbol{\mu}\boldsymbol{\mu}^{\mathrm{T}}. \tag{1.46}$$

Note that the diagonal elements of Γ are the variances of the individual variables.

The characteristic function for this multivariate Gaussian is

$$\chi(\mathbf{s}) = \int_{-\infty}^{\infty} P(\mathbf{x}) \exp(i\mathbf{s} \cdot \mathbf{x}) \, dx_1 \ldots dx_N$$

$$= \exp\left(-\mathbf{s}^{\mathrm{T}} \Gamma \mathbf{s}\right) \exp(i\mathbf{s} \cdot \boldsymbol{\mu}), \tag{1.47}$$

where $\mathbf{s} \equiv (s_1, \ldots, s_N)$.

It is also useful to know that all the higher moments of a Gaussian can be written in terms of the means and covariances. Defining $\Delta X \equiv X - \langle X \rangle$, for a one-dimensional Gaussian we have

$$\langle \Delta X^{2n} \rangle = \frac{(2n-1)!(V_X)^n}{2^{n-1}(n-1)!},$$ (1.48)

$$\langle \Delta X^{2n-1} \rangle = 0,$$ (1.49)

for $n = 1, 2, \ldots$.

Further reading

A beautiful account of probability theory and its applications in inference, measurement, and estimation (all being essentially the same thing), is given in *Probability Theory: The Logic of Science* by E. T. Jaynes [1]. We also recommend the collection of Jaynes' works on the subject, entitled *E. T. Jaynes: Papers on Probability, Statistics, and Statistical Physics* [2]. Both Fourier transforms and distributions (such as the δ-function, also known as the "unit impulse") are discussed in most textbooks on signal processing. See for example the text *Linear Systems* by Sze Tan [3], and *Signals and Systems* by Alan Oppenheim and Alan Willsky [4]. A nice introduction to the theory of distributions is given in *The Theory of Distributions: A Nontechnical Introduction* by Ian Richards and Heekyung Youn [5]. The application of probability to information theory may be found in Shannon's classic book, *The Mathematical Theory of Communication* [6].

Exercises

1. From the joint probability given in Eq. (1.12), calculate the marginal probability density for X, $P(x)$. Also, calculate the conditional probability for Y given X, $P(y|x)$.
2. From the joint probability given in Eq. (1.12), calculate $\langle XY \rangle$. From the marginals of $P(x, y)$, $P(x)$ and $P(y)$, obtain the expectation values $\langle X \rangle$ and $\langle Y \rangle$.
3. If $Y = aX + b$, show that $\langle Y \rangle = a\langle X \rangle + b$ and $V_Y = a^2 V_X$.
4. If X is a random variable with the Gaussian probability density

$$P(x) = \frac{1}{\sqrt{2\pi\sigma^2}} e^{-\frac{x^2}{2\sigma^2}}$$ (1.50)

then what is the probability density of $Z = X^2$?
5. (i) Calculate the characteristic function of the Gaussian probability density given in Eq. (1.5). (ii) Use this characteristic function to calculate the probability density of the sum of two random variables each with this Gaussian density. Hint: you don't have to calculate the inverse transform from the definition; you work it out directly from the answer to (i).

6. Calculate the characteristic function for the probability density

$$P(x) = \frac{1}{\sqrt{8\pi a x}} e^{-\frac{x}{2a}}, \tag{1.51}$$

where x takes values in the interval $(0, \infty)$.

7. The random variables X and Y are independent, and both have a Gaussian probability density with zero mean and unit variance. (i) What is the joint probability density for X and $Z = X + Y$? (ii) What is the conditional probability density for Y given Z? Hint: use the fact that $P(y|z) = P(z|y)P(y)/P(z)$. (Note: this relationship is called *Bayes' theorem*, and is the cornerstone of measurement theory, also known as statistical inference [1].)

8. The random variable X can take any value in the range $(0, \infty)$. Find a probability density for X such that the probability density for $Y = \alpha X$, where α is any positive number, is the same as the density for X. Note that the probability density you get is *not normalizable* (that is, its integral is infinite). This probability density is *scale invariant*, and has uses in statistical inference, even though it is not normalizable [1].

9. The two random variables X and Y have a joint probability density such that the point (X, Y) is uniformly distributed on the unit circle.

(i) What is the joint probability density for X and Y?
(ii) Calculate $\langle X \rangle$, $\langle Y \rangle$ and $\langle XY \rangle$.
(iii) Are X and Y independent?

2

Differential equations

2.1 Introduction

A differential equation is an equation that involves one or more of the derivatives of a function. Let us consider a simple physical example. Say we have a toy train on a straight track, and x is the position of the train along the track. If the train is moving then x will be a function of time, and so we write it as $x(t)$. If we apply a constant force of magnitude F to the train, then its acceleration, being the second derivative of x, is equal to F/m, where m is the train's mass. Thus we have the simple *differential equation*

$$\frac{d^2x}{dt^2} = \frac{F}{m}.$$ (2.1)

To find how x varies with time, we need to find the function $x(t)$ that satisfies this equation. In this case it is very simple, since we can just integrate both sides of the equation twice with respect to t. This gives

$$x(t) = Ft^2/2 + at + b,$$ (2.2)

where a and b are the constants of integration. These constants are determined by the initial value of x, which we will denote by $x_0 \equiv x(0)$, and the initial value of dx/dt, which we denote by v_0. To determine b from x_0 and v_0, one sets $t = 0$ in Eq. (2.2). To determine a one differentiates both sides of Eq. (2.2) with respect to time, and then sets $t = 0$. The resulting solution is $x(t) = Ft^2/2 + v_0 t + x_0$.

Now let's take a more non-trivial example. Let's say we have a metal ball hanging on the end of a spring. If we call the equilibrium position of the ball $x = 0$, then the force on the ball is equal to $-kx$, where k is a positive constant and x is the vertical deviation of the ball from the equilibrium position. Thus the differential equation describing the motion of the ball (often called the *equation of motion*) is

$$\frac{d^2x}{dt^2} = -\left(\frac{k}{m}\right)x.$$ (2.3)

We will show how to solve equations like this later. For now we merely note that one solution is $x(t) = x(0)\cos(\omega t)$, where $\omega = \sqrt{k/m}$. This shows that the ball oscillates up and down after being positioned away from the equilibrium point (in this case by a distance $x(0)$). Both Eq. (2.1) and Eq. (2.3) are called *second-order* differential equations, because they contain the second derivative of x.

2.2 Vector differential equations

We can change a second-order differential equation for some variable x into a set of two differential equations that only contain first derivatives. To do this we introduce a second variable, and set this equal to the first derivative of x. Using as our example the differential equation for the ball on the spring, Eq. (2.3), we now define $p(t) = mdx/dt$, where m is the mass of the ball (so that $p(t)$ is the momentum of the ball). We now have two first-order differential equations for the two functions $x(t)$ and $p(t)$:

$$\frac{dx}{dt} = \frac{p}{m} \quad \text{and} \quad \frac{dp}{dt} = -kx. \tag{2.4}$$

The differential equation for x is simply the definition of p, and the differential equation for p is obtained by substituting the definition of p into the original second-order differential equation for x (Eq. (2.3)).

We can now write this set of first-order differential equations in the "vector form",

$$\frac{d}{dt}\begin{pmatrix} x \\ p \end{pmatrix} \equiv \begin{pmatrix} dx/dt \\ dp/dt \end{pmatrix} = \begin{pmatrix} p/m \\ -kx \end{pmatrix} = \begin{pmatrix} 0 & 1/m \\ -k & 0 \end{pmatrix}\begin{pmatrix} x \\ p \end{pmatrix}. \tag{2.5}$$

Defining $\mathbf{x} = (x, p)^{\mathrm{T}}$ and A as the matrix

$$A = \begin{pmatrix} 0 & 1/m \\ -k & 0 \end{pmatrix}, \tag{2.6}$$

we can write the set of equations in the compact form

$$\dot{\mathbf{x}} \equiv \frac{d\mathbf{x}}{dt} = A\mathbf{x}. \tag{2.7}$$

If the elements of the matrix A do not depend on $\mathbf{x}$, as in the equation above, then this differential equation is referred to as a *linear* first-order vector differential equation.

We can use a similar procedure to transform any nth-order differential equation for x into a set of n first-order differential equations. In this case one defines n variables, $x_1, \ldots, x_n$, with $x_1 \equiv x$, and $x_m \equiv dx_{m-1}/dt$, for $m = 2, \ldots, n$. The definitions of the variables x_2 to x_n give us $n - 1$ differential equations, and the final differential equation, being the equation for $dx_n/dt = d^n x/dt^n$, is given by substituting x_n into the original nth-order differential equation for x.

2.3 Writing differential equations using differentials

We now introduce another way of writing differential equations, as this will be most useful for the stochastic differential equations that we will encounter later. Instead of focussing on the derivative of x at each time t, we will instead consider the *change* in x in an infinitesimal time-step dt. We will call this change dx. By *infinitesimal* we mean a time-step that is small enough that only the first derivative of x contributes significantly to the change that x experiences in that interval. The change dx is given by

$$dx = \frac{dx}{dt}dt. \tag{2.8}$$

We can write our differential equations in terms of dx and dt instead of using the derivatives as we did above. Thus the differential equation given by Eq. (2.5) can be written instead as

$$d\begin{pmatrix} x \\ p \end{pmatrix} \equiv \begin{pmatrix} dx \\ dp \end{pmatrix} = \frac{1}{m}\begin{pmatrix} pdt \\ -kxdt \end{pmatrix} = \frac{1}{m}\begin{pmatrix} 0 & 1 \\ -k & 0 \end{pmatrix}\begin{pmatrix} x \\ p \end{pmatrix}dt, \tag{2.9}$$

or in the more compact form

$$\mathbf{dx} = A\mathbf{x}dt, \tag{2.10}$$

where $\mathbf{x} = (x, p)^\mathrm{T}$, and A is as defined in Eq. (2.6). The infinitesimal increments dx, dt, etc., are called "differentials", and so writing differential equations in this way is often referred to as writing them in "differential form".

2.4 Two methods for solving differential equations

The usual method of solving a first-order differential equation of the form

$$\frac{dx}{dt} = g(t)f(x) \tag{2.11}$$

is to divide by $f(x)$, multiply by dt, and then integrate both sides:

$$\int \frac{dx}{f(x)} = \int g(t)dt + C, \tag{2.12}$$

where C is the constant of integration. This constant is determined by the initial condition after one has performed the integration and solved the resulting equation for x. This method is called "separation of variables", because it works by separating out all the dependence on x to one side, and the dependence on t to the other. It works for linear first-order equations, as well as many nonlinear first-order equations. (An equation is linear if it contains the variables and their derivatives, but not higher powers or any more complex functions of these things.) We will need to use the above method to solve nonlinear equations in later chapters, but for now we are concerned only with the linear case.

We now present an alternative method for solving linear differential equations, because this will be useful when we come to solving stochastic equations, and helps to get us used to thinking in terms of differentials. Let's say we have the simple linear differential equation

$$dx = -\gamma x dt. \tag{2.13}$$

This tells us that the value of x at time $t + dt$ is the value at time t plus dx. That is

$$x(t + dt) = x(t) - \gamma x(t)dt = (1 - \gamma dt)x(t). \tag{2.14}$$

To solve this we note that to first order in dt (that is, when dt is very small) $e^{\alpha dt} \approx 1 + \alpha dt$. We can therefore write the equation for $x(t + dt)$ as

$$x(t + dt) = x(t) - \gamma x(t)dt = e^{-\gamma dt}x(t). \tag{2.15}$$

This tells us that to move x from time t to $t + dt$ we merely have to multiply $x(t)$ by the factor $e^{-\gamma dt}$. So to move by two lots of dt we simply multiply by this factor twice:

$$x(t + 2dt) = e^{-\gamma dt}x(t + dt) = e^{-\gamma dt}\left[e^{-\gamma dt}x(t)\right] = e^{-\gamma 2dt}x(t). \tag{2.16}$$

To obtain $x(t + \tau)$ all we have to do is apply this relation repeatedly. Let us say that $dt = \tau/N$ for N as large as we want. Thus dt is a small but finite time-step, and we can make it as small as we want. That means that to evolve x from time t to $t + \tau$ we have to apply Eq. (2.15) N times. Thus

$$x(t + \tau) = (e^{-\gamma dt})^N x(t) = e^{-\gamma \sum_{n=1}^{N} dt}x(t) = e^{-\gamma Ndt}x(t) = e^{-\gamma \tau}x(t) \tag{2.17}$$

is the solution to the differential equation. If γ is a function of time, so that the equation becomes

$$dx = -\gamma(t)x dt, \tag{2.18}$$

we can still use the above technique. As before we set $dt = \tau/N$ so that it is a small finite time-step. But this time we have to explicitly take the limit as $N \to \infty$

to obtain the solution to the differential equation:

$$x(t + \tau) = \lim_{N \to \infty} \Pi_{n=1}^{N} e^{-\gamma(t+ndt)dt} x(t)$$

$$= \lim_{N \to \infty} e^{-\sum_{n=1}^{N} \gamma(t+ndt)dt} x(t)$$

$$= e^{-\int_{t}^{t+\tau} \gamma(t)dt} x(t). \tag{2.19}$$

The equation we have just solved is the simplest *linear* differential equation. All linear differential equations can be written in the form of Eq. (2.10) above, where the matrix A is independent of $\mathbf{x}$.

2.4.1 A linear differential equation with driving

We will shortly show how one solves linear differential equations when there is more than one variable. But before we do, we consider the simple linear differential equation given by Eq. (2.13) with the addition of a *driving term*. A driving term is a function of time that is independent of the variable. So the single-variable linear equation with driving is

$$\frac{dx}{dt} = -\gamma x + f(t), \tag{2.20}$$

where f is any function of time. To solve this we first transform to a new variable, $y(t)$, defined as

$$y(t) = x(t)e^{\gamma t}. \tag{2.21}$$

Note that $y(t)$ is defined precisely so that if $x(t)$ was a solution to Eq. (2.13), then y would be constant. Now we calculate the differential equation for y. This is

$$\frac{dy}{dt} = \left(\frac{\partial y}{\partial x}\right)\frac{dx}{dt} + \frac{\partial y}{\partial t} = e^{\gamma t} f(t). \tag{2.22}$$

The equation for y is solved merely by integrating both sides, and the solution is

$$y(t) = y_0 + \int_0^t e^{\gamma s} f(s)ds, \tag{2.23}$$

where we have defined y_0 as the value of y at time $t = 0$. Now we can easily obtain $x(t)$ from $y(t)$ by inverting Eq. (2.21). This gives us the solution to Eq. (2.20), which is

$$x(t) = x_0 e^{-\gamma t} + e^{-\gamma t} \int_0^t e^{\gamma s} f(s)ds = x_0 e^{-\gamma t} + \int_0^t e^{-\gamma(t-s)} f(s)ds. \tag{2.24}$$

We can just as easily solve a linear equation when the coefficient γ is a function of time. In this case we transform to $y(t) = x(t) \exp[\Gamma(t)]$, where

$$\Gamma(t) \equiv \int_0^t \gamma(s)ds, \tag{2.25}$$

and the solution is

$$x(t) = x_0 e^{-\Gamma(t)} + e^{-\Gamma(t)} \int_0^t e^{\Gamma(s)} f(s)ds. \tag{2.26}$$

2.5 Solving vector linear differential equations

We can usually solve a linear equation with more that one variable,

$$\dot{\mathbf{x}} = A\mathbf{x}, \tag{2.27}$$

by transforming to a new set of variables, $\mathbf{y} = U\mathbf{x}$, where U is a matrix chosen so that the equations for the new variables are *decoupled* from each other. That is, the equation for the vector $\mathbf{y}$ is

$$\dot{\mathbf{y}} = D\mathbf{y}, \tag{2.28}$$

where D is a diagonal matrix. For many square matrices A, there exists a matrix U so that D is diagonal. This is only actually guaranteed if A is normal, which means that $A^\dagger A = AA^\dagger$. Here $A^\dagger$ is the *Hermitian conjugate* of A, defined as the transpose of the complex conjugate of A. If there is no U that gives a diagonal D, then one must solve the differential equation using Laplace transforms instead, a method that is described in most textbooks on differential equations (for example [7]). If U exists then it is *unitary*, which means that $U^\dagger U = UU^\dagger = I$. The diagonal elements of D are called the *eigenvalues* of A.

There are systematic numerical methods to find the U and corresponding D for a given A, and numerical software such as Matlab and Mathematica include routines to do this. If A is two-by-two or three-by-three, then one can calculate U and D analytically, and we show how this is done in the next section.

If D is diagonal, so that

$$D = \begin{pmatrix} \lambda_1 & 0 & \cdots & 0 \\ 0 & \lambda_2 & \cdots & 0 \\ \vdots & \vdots & \ddots & 0 \\ 0 & 0 & 0 & \lambda_N \end{pmatrix}, \tag{2.29}$$

then for each element of y (each variable), y_n, we have the simple equation

$$\dot{y}_n = \lambda_n y_n, \tag{2.30}$$

and this has the solution $y_n(t) = y_n(0)e^{\lambda_n t}$. The solution for **y** is thus

$$
\mathbf{y}(t) = \begin{pmatrix} e^{\lambda_1 t} & 0 & \cdots & 0 \\ 0 & e^{\lambda_2 t} & \cdots & 0 \\ \vdots & \vdots & \ddots & 0 \\ 0 & 0 & 0 & e^{\lambda_N t} \end{pmatrix} \mathbf{y}(0) \equiv e^{Dt}\mathbf{y}(0), \tag{2.31}
$$

where we have defined the exponential of a diagonal matrix, e^{Dt}.

To get the solution for $\mathbf{x}(t)$, we use the fact that $U^\dagger U = I$, from which it follows immediately that $\mathbf{x} = U^\dagger \mathbf{y}$. Substituting this, along with the definition of **y** into the solution for **y**, we get

$$
\mathbf{x}(t) = U^\dagger e^{Dt} U \mathbf{x}(0). \tag{2.32}
$$

Further, it makes sense to define the exponential of any square matrix A as

$$
e^{At} = U^\dagger e^{Dt} U. \tag{2.33}
$$

To see why, first note that, by substituting **x** into the differential equation for **y**, we find that

$$
\dot{\mathbf{x}} = U^\dagger D U \mathbf{x}, \tag{2.34}
$$

and thus $At = U^\dagger DtU$. Because of this, the power series

$$
\sum_{n=0}^{\infty} \frac{(At)^n}{n!} = 1 + At + \frac{(At)^2}{2} + \frac{(At)^3}{3!} + \cdots
$$

$$
= 1 + U^\dagger DtU + \frac{U^\dagger DtU U^\dagger DtU}{2} + \frac{(U^\dagger DtU)^3}{3!} + \cdots
$$

$$
= 1 + U^\dagger DtU + \frac{U^\dagger (Dt)^2 U}{2} + \frac{U^\dagger (Dt)^3 U}{3!} + \cdots
$$

$$
= U^\dagger \left(1 + Dt + \frac{(Dt)^2}{2} + \frac{(Dt)^3}{3!} + \cdots \right) U
$$

$$
= U^\dagger e^{Dt} U. \tag{2.35}
$$

So $U^\dagger e^{Dt} U$ corresponds precisely to the power series $\sum_{n}^{\infty}(At)^n/n!$, which is the natural generalization of the exponential function for a matrix At. Since the above relationship holds for all power series, the natural definition of *any* function of a square matrix A is

$$
f(A) \equiv U^\dagger f(D)U \equiv U^\dagger \begin{pmatrix} f(\lambda_1) & 0 & \cdots & 0 \\ 0 & f(\lambda_2) & \cdots & 0 \\ \vdots & \vdots & \ddots & 0 \\ 0 & 0 & 0 & f(\lambda_N) \end{pmatrix} U. \tag{2.36}
$$

To summarize the above results, the solution to the vector differential equation

$$\dot{\mathbf{x}} = A\mathbf{x}, \tag{2.37}$$

is

$$\mathbf{x}(t) = e^{At}\mathbf{x}(0), \tag{2.38}$$

where

$$e^{At} \equiv \sum_{n}^{\infty} \frac{(At)^n}{n!} = U^\dagger e^{Dt} U. \tag{2.39}$$

We can now also solve any linear vector differential equation with driving, just as we did for the single-variable linear equation above. The solution to

$$\dot{\mathbf{x}} = A\mathbf{x} + \mathbf{f}(t), \tag{2.40}$$

where $\mathbf{f}$ is now a vector of driving terms, is

$$\mathbf{x}(t) = e^{At}\mathbf{x}(0) + \int_0^t e^{A(t-s)}\mathbf{f}(s)ds. \tag{2.41}$$

2.6 Diagonalizing a matrix

To complete this chapter, we now show how to obtain the matrices U and D for a square N dimensional (N by N) matrix A. It is feasible to use this to obtain analytical expressions for U and D when A is two dimensional, and for three dimensions if A has a sufficiently simple form.

First we need to find a set of N special vectors, called the *eigenvectors* of A. An eigenvector is a vector, $\mathbf{v}$, for which

$$A\mathbf{v} = \lambda\mathbf{v}, \tag{2.42}$$

where λ is a number (real or complex). To find all the eigenvectors we note that if Eq. (2.42) is true, then

$$(A - \lambda I)\mathbf{v} = 0, \tag{2.43}$$

where I is the N-dimensional identity matrix. This is true if and only if the determinant of $A - \lambda I$ is zero. The equation

$$\det |A - \lambda I| = 0 \tag{2.44}$$

is an Nth-order polynomial equation for λ. This has N solutions, giving in general N different eigenvalues. Note that some of the eigenvalues may be the same, because the polynomial may have repeated roots. In this case there are fewer than

N distinct eigenvalues, but so long as A is Hermitian symmetric there will always be N distinct (and orthogonal) eigenvectors; each of the repeated roots will have more than one corresponding eigenvector. For $N = 2$ and $N = 3$ there exist analytical expressions for the roots of the polynomial, and thus for the eigenvalues. We will denote the N eigenvalues as λ_i, $i = 1, \ldots, N$ (some of which may be repeated), and the corresponding eigenvectors as $\mathbf{v}_i$.

For each distinct (unrepeated) eigenvalue, λ_i, one determines the corresponding eigenvector by solving the equation

$$A\mathbf{v}_i = \lambda_i \mathbf{v}_i, \tag{2.45}$$

for $\mathbf{v}_i$. For an eigenvalue that is repeated m times (also known as a *degenerate* eigenvalue), we solve the same equation, except that now the solution is an m-dimensional vector space. In this case we then choose m-linearly independent vectors in this space. On obtaining these vectors, the Gramm–Schmidt [8] orthogonalization procedure can then be used to obtain m mutually orthogonal vectors that span the space. (If you are not familiar with vector spaces and the associated terminology, then this information can be obtained from a textbook on linear algebra [7].)

Having followed the above procedure, we now have N mutually orthogonal eigenvectors, $\mathbf{v}_n$, each with a corresponding eigenvalue, λ_n. All that remains to be done is to divide each eigenvector by its norm, $|\mathbf{v}_n| = \sqrt{\mathbf{v}_n \cdot \mathbf{v}_n}$, which gives us a set of ortho*normal* eigenvectors, $\tilde{\mathbf{v}}_n = \mathbf{v}_n/|\mathbf{v}_n|$. Defining $U^\dagger$ by

$$U^\dagger = (\tilde{\mathbf{v}}_1 \tilde{\mathbf{v}}_2 \ldots \tilde{\mathbf{v}}_N) \tag{2.46}$$

(so that the columns of $U^\dagger$ are the orthonormal eigenvectors), and D by Eq. (2.29), it is straightforward to verify that $UU^\dagger = I$ and that $D = UAU^\dagger$. It is also true, though not as obvious, that $U^\dagger U = I$.

Exercises

1. The equation for the damped harmonic oscillator is:

$$m\frac{d^2x}{dt^2} + 2\gamma m\frac{dx}{dt} + kx = 0. \tag{2.47}$$

Assume that $\gamma < \sqrt{k/m}$, in the which the resonator is said to be "under-damped". Show that $x(t) = e^{-\gamma t}\cos(\omega t)x(0)$ is a solution to Eq. (2.47), and find the expression for the frequency ω in terms of m, k and γ.

2. Write Eq. (2.47) as a first-order vector differential equation.

3. Solve the linear vector differential equation

$$\dot{\mathbf{x}} = A\mathbf{x} \tag{2.48}$$

with

$$A = \begin{pmatrix} 0 & \omega \\ -\omega & -\gamma \end{pmatrix}. \tag{2.49}$$

Do this by calculating the eigenvectors and eigenvalues of the matrix A. Assume that $\gamma < \omega$.

4. Use the answer to Exercise **3** above to write down the solution to the differential equation

$$\frac{d^2x}{dt^2} + \gamma \frac{dx}{dt} + \omega^2 x = f(t). \tag{2.50}$$

5. Calculate the eigenvalues and eigenvectors of the matrix

$$A = \begin{pmatrix} 2 & 0 & \sqrt{\frac{2}{3}} \\ 0 & 2 & \frac{-1}{\sqrt{3}} \\ \sqrt{\frac{2}{3}} & \frac{-1}{\sqrt{3}} & 2 \end{pmatrix} \tag{2.51}$$

and use them to solve the differential equation

$$\dot{\mathbf{x}} = A\mathbf{x}. \tag{2.52}$$

6. Use the power series for the exponential function to show that

$$e^{\lambda \sigma} = \cosh(\lambda)I + \sinh(\lambda)\sigma, \tag{2.53}$$

where

$$\sigma = \begin{pmatrix} 0 & 1 \\ 1 & 0 \end{pmatrix} \tag{2.54}$$

and I is the two-by-two identity matrix.

3

Stochastic equations with Gaussian noise

3.1 Introduction

We have seen that a differential equation for x is an equation that tells us how x changes in each infinitesimal time-step dt. In the differential equations we have considered so far, this increment in x is deterministic – that is, it is completely determined at each time by the differential equation. Now we are going to consider a situation in which this increment is not completely determined, but is a random variable. This is the subject of *stochastic differential equations*.

It will make things clearest if we begin by considering an equation in which time is divided into finite (that is, not infinitesimal) chunks of duration Δt. In this case x gets a finite increment, Δx, during each time-chunk. Such an equation is referred to as a *discrete-time* equation, or a *difference* equation, because it specifies the difference, Δx, between $x(t)$ and $x(t + \Delta t)$. Since time moves in steps of Δt, x takes values only at the set of times $t_n = n\Delta t$, where $n = 0, 1, \ldots, \infty$.

Let us consider a discrete-time equation in which the change in x, Δx, is the sum of a term that is proportional to x, and a "driving term" that is independent of x. We will make this driving term a function of time, thus giving it the freedom to be different at different times. The difference equation for x is

$$\Delta x(t_n) = x(t_n)\Delta t + f(t_n)\Delta t. \tag{3.1}$$

where $f(t_n)\Delta t$ is the driving term.

Given the value of x at time t_n, the value at time t_{n+1} is then

$$x(t_n + \Delta t) = x(t_n) + \Delta x(t_n) = x(t_n) + x(t_n)\Delta t + f(t_n)\Delta t. \tag{3.2}$$

If we know the value of x at $t = 0$, then

$$x(\Delta t) = x(0)(1 + \Delta t) + f(0)\Delta t. \tag{3.3}$$

26

Now, what we are really interested in is what happens if the driving term, $f(t_n)\Delta t$ is *random* at each time t_n? This means replacing $f(t_n)$ with a random variable, y_n, at each time t_n. Now the difference equation for x becomes

$$\Delta x(t_n) = x(t_n)\Delta t + y_n \Delta t, \tag{3.4}$$

and this is called a "stochastic difference equation". This equation says that at each time t_n we pick a value for the random variable y_n (sampled from its probability density), and add $y_n \Delta t$ to $x(t_n)$. This means that we can no longer predict exactly what x will be at some future time, T, until we arrive at that time, and all the values of the random increments up until T have been determined.

The solution for x at time Δt is

$$x(\Delta t) = x(0)(1 + \Delta t) + y_0 \Delta t. \tag{3.5}$$

So $x(\Delta t)$ is now a random variable. If $x(0)$ is fixed (that is, not random), then $x(\Delta t)$ is just a linear transformation of the random variable y_0. If $x(0)$ is also random, then $x(\Delta t)$ is a linear combination of the two random variables $x(0)$ and y_0. Similarly, if we go to the next time-step, and calculate $x(2\Delta t)$, then this is a function of $x(0)$, y_0 and y_1. We see that at each time then, the solution to a stochastic difference equation, $x(t_n)$, is a random variable, and this random variable changes at each time-step. To solve a stochastic difference equation, we must therefore determine the *probability density* for x at all future times. Since $x(t_n)$ is a function of all the noise increments y_n, as well as $x(0)$, this means calculating the probability density for x from the probability densities for the noise increments (and from the probability density for $x(0)$ if it is also random). That is why we discussed summing random variables, and making transformations of random variables in Chapter 1.

Stochastic *differential* equations are obtained by taking the limit $\Delta t \to 0$ of stochastic difference equations. Thus the solution of a stochastic differential equation is also a probability density for the value of x at each future time t. Just as in the case of ordinary (deterministic) differential equations, it is not always possible to find a closed-form expression for the solution of a stochastic differential equation. But for a number of simple cases it is possible. For stochastic equations that cannot be solved analytically, we can solve them numerically using a computer, and we describe this later in Chapter 6.

The reason that differential equations that are driven by random increments at each time-step are called *stochastic* differential equations is because a process that fluctuates randomly in time is called a *stochastic process*. Stochastic differential equations (SDEs) are thus driven by stochastic processes. The random increments that drive an SDE are also referred to as *noise*.

In addition to obtaining the probability density for x at the future times t_n, we can also ask how x evolves with time given a specific set of values for the random increments y_n. A set of values of the random increments (values sampled from their probability densities) is called a *realization* of the noise. A particular evolution for x given a specific noise realization is called a *sample path* for x. In addition to wanting to know the probability density for x at future times, it can also be useful to know what properties the sample paths of x have. The full solution to an SDE is therefore really the complete set of all possible sample paths, and the probabilities for all these paths, but we don't usually need to know all this information. Usually all we need to know is the probability density for x at each time, and how x at one time is correlated with x at another time. How we calculate the latter will be discussed in Chapter 4. Note that since the solution to an SDE, $x(t)$, varies randomly in time, it is a stochastic process. If one wishes, one can therefore view an SDE as something that takes as an input one stochastic process (the driving noise), and produces another.

3.2 Gaussian increments and the continuum limit

In this chapter we are going to study stochastic differential equations driven by Gaussian noise. By "Gaussian noise" we mean that each of the random increments has a Gaussian probability density. First we consider the simplest stochastic difference equation, in which the increment of x, Δx, consists solely of the random increment $y_n \Delta t$. Since Gaussian noise is usually called *Wiener noise*, we will call the random increment $\Delta W_n = y_n \Delta t$. The discrete differential equation for x is thus

$$\Delta x(t_n) = \Delta W_n. \tag{3.6}$$

Each *Wiener* increment is completely independent of all the others, and all have the same probability density, given by

$$P(\Delta W) = \frac{1}{\sqrt{2\pi\sigma^2}} e^{-(\Delta W)^2/(2\sigma^2)}. \tag{3.7}$$

This density is a Gaussian with zero mean, and we choose the variance to be $\sigma^2 = V = \Delta t$. This choice for the variance of the Wiener increment is very important, and we will see why shortly. We will often denote a Wiener increment in some time-step Δt simply as ΔW, without reference to the subscript n. This notation is convenient because all the random increments have the same probability density, and even though independent of each other, are in that sense identical.

We can easily solve the difference equation for x simply by starting with $x(0) = 0$, and repeatedly adding Δx. This gives the solution

$$x_n \equiv x(n\Delta t) = \sum_{i=0}^{n-1} \Delta W_i. \tag{3.8}$$

So now we need to calculate the probability density for x_n. Since we know from Exercise 5 in Chapter 1 that the sum of two Gaussian random variables is also a Gaussian, we know that the probability density for x_n is Gaussian. We also know that the mean and variance of x_n is the sum of the means and variances of the ΔW_i, because the ΔW_i are all independent (see Section 1.5). We therefore have

$$\langle x_n \rangle = 0, \tag{3.9}$$

$$V(x_n) = n\Delta t, \tag{3.10}$$

and so

$$P(x_n) = \frac{1}{\sqrt{2\pi V}} e^{-x_n^2/(2V)} = \frac{1}{\sqrt{2\pi n \Delta t}} e^{-x_n^2/(2n\Delta t)}. \tag{3.11}$$

We now need to move from difference equations to *differential* equations. To do so we will consider solving the difference equation above, Eq. (3.6), for a given future time T, with N discrete time-steps. (So each time-step is $\Delta t = T/N$.) We then take the limit as $N \to \infty$. Proceeding in the same way as above, the solution $x(T)$ is now

$$x(T) = \lim_{N\to\infty} \sum_{i=0}^{N-1} \Delta W_i \equiv \int_0^T dW(t) \equiv W(T). \tag{3.12}$$

Here we define a *stochastic integral*, $W(T) = \int_0^T dW(t)$, as the limit of the sum of all the increments of the Wiener process. A stochastic integral, being the sum of a bunch of random variables, is therefore a random variable. The important thing is that in many cases we can calculate the probability density for this random variable. We can easily do this in the case above because we know that the probability density for $x(T)$ is Gaussian (since it is merely the sum of many independent Gaussian variables). Because of this, we need only to calculate its mean and variance. The mean is zero, because all the random variables in the sum have zero mean. To calculate the variance of $x(T)$ it turns out that we don't even need to take the limit as $N \to \infty$ because N factors out of the expression:

$$V(x(T)) = \sum_{i=0}^{N-1} V[\Delta W_i] = \sum_{i=0}^{N-1} \Delta t = N\Delta t = N\left(\frac{T}{N}\right) = T. \tag{3.13}$$

The probability density for $W(T)$ is therefore

$$P(W(T)) = P(x(T)) \equiv P(x, T) = \frac{1}{\sqrt{2\pi T}} e^{-x^2/(2T)}. \tag{3.14}$$

Note that when writing the integral over the Wiener increments, we have included t explicitly as an argument to dW, just to indicate that dW changes with time. We will often drop the explicit dependence of dW on t, and just write the stochastic integral as

$$W(T) = \int_0^T dW. \tag{3.15}$$

While we often loosely refer to the increments dW as the "Wiener process", the Wiener process is actually defined as $W(T)$, and dW is, strictly speaking, an increment of the Wiener process.

The fact that the variance of $x(T)$ is proportional to T is a result of the fact that we chose each of the Wiener increments to have variance Δt. Since there is one Wiener increment in each time step Δt, the variance of x grows by precisely Δt in each interval Δt, and is thus proportional to t. So what would happen if we chose $V[\Delta W(\Delta t)]$ to be some other power of Δt? To find this out we set $V[\Delta W(\Delta t)] = \Delta t^\alpha$ and calculate once again the variance of $x(T)$ (before taking the limit as $N \to \infty$). This gives

$$V(x(T)) = \sum_{i=0}^{N-1} V[\Delta W_i] = N(\Delta t)^\alpha = N \left(\frac{T}{N} \right)^\alpha = N^{(1-\alpha)} T^\alpha. \tag{3.16}$$

Now we take the continuum limit $N \to \infty$ so as to obtain a stochastic differential equation. When α is greater than one we have

$$\lim_{N \to \infty} V(x(T)) = T^\alpha \lim_{N \to \infty} N^{(1-\alpha)} = 0, \tag{3.17}$$

and when α is less than one we have

$$\lim_{N \to \infty} V(x(T)) = T^\alpha \lim_{N \to \infty} N^{(1-\alpha)} \to \infty. \tag{3.18}$$

Neither of these make sense for the purposes of obtaining a stochastic differential equation that describes real systems driven by noise. Thus we are *forced* to choose $\alpha = 1$ and hence $V(\Delta W_n) \propto \Delta t$.

When we are working in the continuum limit the Gaussian increments, dW, are referred to as being *infinitesimal*. A general SDE for a single variable $x(t)$ is then written as

$$dx = f(x, t)dt + g(x, t)dW. \tag{3.19}$$

Since the variance of dW must be proportional to dt, and since any constant of proportionality can always be absorbed into $g(x, t)$, the variance of dW is defined to be *equal* to dt. We can therefore write the probability density for dW as

$$P(dW) = \frac{e^{-(dW)^2/(2dt)}}{\sqrt{2\pi dt}}. \tag{3.20}$$

To summarize the main result of this section: if we have an SDE driven by an infinitesimal increment dW in each infinitesimal interval dt, then if all the increments are **Gaussian**, **independent** and **identical**, they *must* have variance proportional to dt.

3.3 Interlude: why Gaussian noise?

In this chapter we consider only noise that is Gaussian. This kind of noise is important because it is very common in physical systems. The reason for this is a result known as the *central limit theorem*. The central limit theorem says that if one sums together many independent random variables, then the probability density of the sum will be close to a Gaussian. As the number of variables in the sum tends to infinity, the resulting probability density tends exactly to a Gaussian. The only condition on the random variables is they all have a finite variance.

Now consider noise in a physical system. This noise is usually the result of many random events happening at the microscopic level. This could be the impacts of individual molecules, the electric force from many electrons moving randomly in a conductor, or the thermal jiggling of atoms in a solid. The total force applied by these microscopic particles is the sum of the random forces applied by each. Because the total force is the sum over many random variables, it has a Gaussian probability density. Since the microscopic fluctuations are usually fast compared to the motion of the system, we can model the noise as having a Gaussian probability density in each time-step Δt, where Δt is small compared to the time-scale on which the system moves. In fact, assuming that the noise increments are completely independent of each other from one infinitesimal time-step to the next is really an idealization (an approximation) which is not true in practice. Nevertheless, this approximation works very well, and we will explain why at the end of Section 4.6.

From a mathematical point of view, simple noise processes in which the random increment in each time interval dt is independent of all the previous random increments will usually be Gaussian for the same reason. This is because the random increment in each small but finite time interval Δt is the sum over the infinite number of increments for the infinitesimal intervals dt that make up that finite interval. There are two exceptions to this. One is processes in which the random increment in an infinitesimal time-step dt is not necessarily infinitesimal. The sample paths of such processes make instant and discrete jumps from time to time, and are thus not continuous. These are called *jump* or *point* processes, and we consider them in Chapter 8. Jump processes are actually quite common and have many applications. The other exception, which is much rarer in nature, happens when the noise increments remain infinitesimal, as in

Gaussian noise, but are drawn from a probability density with an infinite variance (one that avoids the central limit theorem). These processes will be discussed in Chapter 9.

3.4 Ito calculus

We saw in the previous section that the increments of Wiener noise, dW, have a variance proportional to dt. In this section we are going to discover a very surprising consequence of this fact, and the most unusual aspect of Wiener noise and stochastic equations. To set the stage, consider how we obtained the solution of the simple differential equation

$$dx = -\gamma x dt. \tag{3.21}$$

We solved this equation in Section 2.4 using the relation $e^{\alpha dt} \approx 1 + \alpha dt$. This relation is true because dt is a differential – that is, it is understood that when calculating the solution, $x(t)$, we will always take the limit in which $dt \to 0$, just as we did in solving the simple stochastic equation in the previous section (in fact, we can regard the use of the symbol dt as a shorthand notation for the fact that this limit will be taken). The approximation $e^{\alpha dt} \approx 1 + \alpha dt$ works because the terms in the power series expansion for $e^{\alpha dt}$ that are second-order or higher in dt (dt^2, dt^3, etc.) will vanish in comparison to dt as $dt \to 0$.

The result of being able to ignore terms that are second-order and higher in the infinitesimal increment leads to the usual rules for differential equations. (It also means that any equation we write in terms of differentials dx and dt can alternatively be written in terms of derivatives.) However, we will now show that the second power of the stochastic differential dW does not vanish with respect to dt, and we must therefore learn a new rule for the manipulation of stochastic differential equations. It will also mean that we cannot write stochastic differential equations in terms of derivatives – we must use differentials.

Solving a differential equation involves summing the infinitesimal increments over all the time-steps dt. To examine whether $(dW)^2$ makes a non-zero contribution to the solution, we must therefore sum $(dW)^2$ over all the time-steps for a finite time T. To do this we will return to a discrete description so that we can explicitly write down the sum and then take the continuum limit.

The first thing to note is that the expectation value of $(\Delta W)^2$ is equal to the variance of ΔW, because $\langle \Delta W \rangle = 0$. Thus $\langle (\Delta W)^2 \rangle = \Delta t$. This tells us immediately that the expectation value of $(\Delta W)^2$ does not vanish with respect to the time-step, Δt, and so the sum of these increments will not vanish when we sum over all the time-steps and take the infinitesimal limit. In fact, the expectation value of the sum of all the increments $(dW)^2$ from 0 to T is simply T:

$$\left\langle \int_0^T (dW)^2 \right\rangle = \int_0^T \langle (dW)^2 \rangle = \int_0^T dt = T. \tag{3.22}$$

Now let us see what the *variance* of the sum of all the $(\Delta W)^2$ is. As we will see shortly, this drops to zero in the continuum limit, so that the integral of all the $(dW)^2$ is not random at all, but deterministic! The reason for this is exactly the same reason that the variance of the average of N independent identical random variables goes to zero as N tends to infinity (see Section 1.5). If we sum N random variables together, but divide each by N so as to keep the mean fixed as N increases, then the variance of each variable drops as $1/N^2$. Because of this the sum of the N variances goes to zero as $N \to \infty$. This is exactly what happens to the sum of all the $(\Delta W)^2$, since the mean of each is equal to $\Delta t = T/N$. The sum of the means remains fixed at the total time interval $T = N\Delta t$, but the sum of the variances drops as $1/N$.

To explicitly calculate the variance of $\int_0^T (dW)^2$, we note that since $\langle (\Delta W)^2 \rangle$ is proportional to Δt, the variance of $(\Delta W)^2$ must be proportional to $(\Delta t)^2$. (We can in fact calculate it directly using the probability density for ΔW, and the result is $V[(\Delta W)^2] = 2(\Delta t)^2$.) So, as discussed above, the variance of $(\Delta W)^2$ is proportional to $1/N^2$:

$$V[(\Delta W)^2] = 2(\Delta t)^2 = 2\frac{T^2}{N^2}. \tag{3.23}$$

The variance of the sum of all the $(\Delta W)^2$ is thus

$$V\left[\sum_{n=0}^{N-1} (\Delta W)^2 \right] = \sum_{n=0}^{N-1} V\left[(\Delta W)^2 \right] = \sum_{n=0}^{N-1} 2(\Delta t)^2 = 2N\left(\frac{T}{N} \right)^2 = \frac{2T^2}{N}. \tag{3.24}$$

And hence the integral of all the $(dW)^2$ is

$$\lim_{N \to \infty} V\left[\sum_{n=0}^{N-1} (\Delta W)^2 \right] = \lim_{N \to \infty} \frac{2T^2}{N} = 0. \tag{3.25}$$

Since the integral of all the $(dW)^2$ is deterministic, it is equal to its mean, T. That is,

$$\int_0^T (dW)^2 = T = \int_0^T dt. \tag{3.26}$$

Thus we have the surprising result that $dW^2 = dt$, a result officially known as Ito's lemma. We will refer to it here as Ito's rule. It is the fundamental rule for solving stochastic differential equations that contain Gaussian noise.

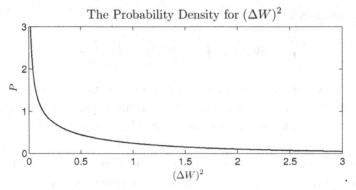

Figure 3.1. The probability density for $(\Delta W)^2$. The mean is $\langle (\Delta W)^2 \rangle = \Delta t$.

If you are not convinced that $(dW)^2 \equiv dt$ merely by calculating the variance of the integral of $(dW)^2$, as we have done above, we can do much better than this. It is not difficult to calculate the entire probability density for the integral of all the $(dW)^2$, and this shows us that the integral is exactly equal to the time interval T. To do this we use the characteristic function. Let us denote the nth square increment, $(\Delta W_n)^2$, as Z_n, and the sum of the square increments as the random variable $Y(T)$:

$$Y(T) = \sum_{n=0}^{N-1} Z_n = \sum_{n=0}^{N-1} [\Delta W_n]^2. \tag{3.27}$$

To calculate the probability density of $Y(T)$, we need to know first the probability density for $Z_n \equiv (\Delta W_n)^2$. Using the probability density for ΔW (Eq. (3.7)) and the method for applying a transformation to a random variable as described in Section 1.6, we obtain

$$P(Z_n) = \frac{e^{-Z_n/(2\Delta t)}}{\sqrt{2\pi \, \Delta t \, Z_n}}. \tag{3.28}$$

This probability density is shown in Figure 3.1. We now take the Fourier transform of this to get the characteristic function, which is

$$\chi_{Z_n}(s) = \frac{1}{\sqrt{1 - 2is\Delta t}} = \frac{1}{\sqrt{1 - 2isT/N}}. \tag{3.29}$$

The characteristic function for $Y(T) = \sum_{n=1}^{N} Z_n$, is then the product of the characteristic functions for each of the Z_n. Hence

$$\chi_Y(s) = \left[\frac{1}{\sqrt{1 - 2isT/N}} \right]^N. \tag{3.30}$$

Now we can quite easily take the continuum limit, which is

$$\chi_Y(s) = \lim_{N \to \infty} \left[\frac{1}{\sqrt{1 - 2isT/N}} \right]^N = \lim_{N \to \infty} \left[1 + \frac{(isT)}{N} \right]^N = e^{isT}. \quad (3.31)$$

In the first step we have used the binomial approximation for the square root,

$$\frac{1}{\sqrt{1-x}} \approx 1 + x/2, \quad (3.32)$$

because N is large. In the second step we have used the definition of the exponential function, $e^x \equiv \lim_{N \to \infty}(1 + x/N)^N$. Finally, we can now obtain the probability density for $Y(T)$ by taking the inverse Fourier transform. This gives

$$P(Y) = \frac{1}{2\pi} \int_{-\infty}^{\infty} e^{-isY} \chi(s) ds = \frac{1}{2\pi} \int_{-\infty}^{\infty} e^{is(T-Y)} ds = \delta(T - Y). \quad (3.33)$$

The probability density for Y at time T is thus a delta function centered at T. Recall from Chapter 1 that a delta function $\delta(x - a)$ is an infinitely sharp function which is zero everywhere except at $x = a$. It is infinitely tall at $x = a$ so that the area underneath it is unity: $\int_{-\infty}^{\infty} \delta(x - a) dx = 1$. The fact that the probability density for Y is a delta function centered at T means that Y is exactly T and has no randomness.

Summary: we have now seen that the result of summing the Wiener increments $(\Delta W)^2$, in the infinitesimal limit, over the time interval T is simply T. This is just what we get if we sum dt over the same time interval. Thus, we have the surprising result that, in the continuum limit,

$$(dW)^2 = dt. \quad (3.34)$$

This relation, which we will refer to as *Ito's rule*, is a departure from the usual rules of calculus. This means that whenever $(dW)^2$ appears in the process of solving a stochastic differential equation (SDE), it cannot be discarded, as can terms that are second and higher order in dt. However, it turns out that all other terms, such as products of the form $dt^n dW^m$, do vanish in the infinitesimal limit. The only terms that ever contribute to the solution of an SDE are dt, dW and $(dW)^2$. Thus Ito's rule is the only additional rule that we will need to know to manipulate SDEs. The calculus of stochastic differential equations (at least in the form we have derived it here) is called *Ito calculus*.

3.5 Ito's formula: changing variables in an SDE

In solving ordinary differential equations in Chapter 2 we were able to ignore terms that were higher than first order in dt. Now that we are working with stochastic differential equations we will have to keep all terms that are first order in dt and

dW, *and* all terms that are second order in dW. In fact, wherever we find terms that are second order in dW we can simply replace them with dt. We need to do this any time we have an SDE for a variable x, and wish to know the resulting SDE for a variable that is some *function* of x. To see how this works, consider a simple example in which we want to know the differential equation for $y = x^2$.

We must first work out the relationship between the increment of y, dy and the increment of x. We have

$$\begin{aligned} dy &\equiv y(t + dt) - y(t) = x(t + dt)^2 - x(t)^2 \\ &= (x + dx)^2 - x^2 \\ &= x^2 + 2x\,dx + (dx)^2 - x^2 \\ &= 2x\,dx + (dx)^2. \end{aligned} \tag{3.35}$$

We see from this example that when we have a nonlinear function of a stochastic variable x, the second power of dx appears in the increment for that function. If x were deterministic then $(dx)^2$ would vanish in the continuum limit, and we would have the usual rule of calculus, being

$$dy = 2x\,dx \quad \text{or} \quad \frac{dy}{dx} = 2x. \tag{3.36}$$

However, when x obeys the stochastic differential equation

$$dx = f\,dt + g\,dW, \tag{3.37}$$

we have

$$\begin{aligned} dy &= 2x\,dx + (dx)^2 \\ &= 2x(f\,dt + g\,dW) + g^2(dW)^2 \\ &= (2fx + g^2)dt + 2xg\,dW. \end{aligned} \tag{3.38}$$

This is Ito's rule in action.

Fortunately there is a simple way to calculate the increment of any nonlinear function $y(x)$ in terms of the first and second powers of the increment of x. All we have to do is use the Taylor series expansion for $y(x)$, truncated at the second term:

$$dy = \left(\frac{dy}{dx}\right)dx + \frac{1}{2}\left(\frac{d^2y}{dx^2}\right)(dx)^2. \tag{3.39}$$

If y is also an explicit function of time as well as x, then this becomes

$$dy = \left(\frac{\partial y}{\partial x}\right)dx + \left(\frac{\partial y}{\partial t}\right)dt + \frac{1}{2}\left(\frac{d^2y}{dx^2}\right)(dx)^2. \tag{3.40}$$

We will refer to Eqs. (3.39) and (3.40) as *Ito's formula*.

3.6 Solving some stochastic equations

3.6.1 The Ornstein–Uhlenbeck process

We now turn to the problem of obtaining analytic solutions to stochastic differential equations. It turns out that there are very few that can be solved analytically, and we will examine essentially all of these in this chapter. The first example is that of a linear differential equation driven by "additive noise". This is

$$dx = -\gamma x dt + g dW. \tag{3.41}$$

The term "additive noise" refers to the fact that the noise does not itself depend on x, but is merely added to any other terms that appear in the equation for dx. This equation is called the *Ornstein–Uhlenbeck equation*, and its solution is called the Ornstein–Uhlenbeck process. To solve Eq. (3.41) we note first that we know the solution to the deterministic part of the equation $dx = -\gamma x dt$, being $x(t) = x_0 e^{-\gamma t}$. We now change variables in the SDE to $y = xe^{\gamma t}$ in the hope that this will simplify the equation. We can use Ito's formula for changing variables, as given in Eq. (3.40), but you may also find it instructive to do it explicitly. The latter method gives

$$
\begin{aligned}
dy &= y(x(t+dt), t+dt) - y(t) \\
&= y(x+dx, t+dt) - y(t) \\
&= (x+dx)e^{\gamma(t+dt)} - xe^{\gamma t} \\
&= xe^{\gamma t}\gamma dt + e^{\gamma t}(1+\gamma dt)dx \\
&= xe^{\gamma t}\gamma dt + e^{\gamma t}dx \\
&= \gamma y dt + e^{\gamma t}dx.
\end{aligned}
\tag{3.42}
$$

In this calculation we have used the fact that the product of any infinitesimal increment with dt is zero. Substituting this expression for dy into the SDE for x gives

$$
\begin{aligned}
dy &= \gamma y dt + e^{\gamma t}dx \\
&= \gamma y dt + e^{\gamma t}(-\gamma x dt + g dW) \\
&= g e^{\gamma t} dW.
\end{aligned}
\tag{3.43}
$$

Now we have an equation that is easy to solve! To do so we merely sum all the stochastic increments dW over a finite time t, noting that each one is multiplied by $ge^{\gamma t}$. The result is

$$y(t) = y_0 + g \int_0^t e^{\gamma s} dW(s). \tag{3.44}$$

This may look rather strange at first. So what *is* this stochastic integral? Well, it is a sum of Gaussian random variables, so it is itself merely a Gaussian random variable. Thus all we need to do is calculate its mean and variance. To do so we return to discrete time-steps so that we can calculate everything explicitly as we have done before. Thus we have

$$y(t) = y(N \Delta t) = y_0 + \lim_{N \to \infty} g \sum_{n=0}^{N-1} e^{\gamma n \Delta t} \Delta W_n, \qquad (3.45)$$

where we have defined $\Delta t = t/N$. The variance of $Y(t)$ is simply the sum of the variances of each of the random variables given by $Y_n = g e^{\gamma n \Delta t} \Delta W_n$. Since multiplying a random variable by a number c changes its variance by a factor of c^2, the variance of Y_n is $g^2 e^{2\gamma n \Delta t} \Delta t$. So we have

$$V[y(t)] = \lim_{N \to \infty} \sum_{n=0}^{N-1} g^2 e^{2\gamma n \Delta t} \Delta t,$$

$$= g^2 \int_0^t e^{2\gamma s} ds = \frac{g^2}{2\gamma} \left(e^{2\gamma t} - 1 \right). \qquad (3.46)$$

Similar reasoning shows that the mean of the stochastic integral in Eq. (3.44) is zero, and so $\langle y(t) \rangle = y_0$. This completely determines $y(t)$.

To obtain the solution of our original differential equation (Eq. (3.41)), we transform back to x using $x = y e^{-\gamma t}$. The solution is

$$x(t) = x_0 e^{-\gamma t} + g e^{-\gamma t} \int_0^t e^{\gamma s} dW(s) = x_0 e^{-\gamma t} + g \int_0^t e^{\gamma(s-t)} dW(s). \qquad (3.47)$$

It is worth noting that the solution we have obtained is exactly the same solution that we would get if we replaced the driving noise by a deterministic function of time, $f(t)$. This would mean replacing dW with $f(t)dt$. The solution has exactly the same form because the method we used above to solve the stochastic equation was exactly the same method that we used in Chapter 2 to solve the same equation but with deterministic driving. That is, to solve this stochastic equation we do not need to use Ito calculus – normal calculus is sufficient. This can be seen immediately by noting that nowhere did $(dW)^2$ appear in the analysis, because $d^2 y/dx^2 = 0$. This is always true if the term giving the driving noise in the equation (the term containing dW) does not contain x. When the driving noise does depend on x, then we cannot obtain the solution by assuming that $dW = f(t)dt$ for some deterministic function $f(t)$, and must use Ito calculus to get the solution. In the next section we show how to solve the simplest equation of this type.

We can also solve the Ornstein–Uhlenbeck stochastic equation when γ and g are functions of time, using essentially the same method. That is, we change to a new variable that has the deterministic part of the dynamics removed, and proceed as before. We leave the details of the derivation as an exercise. The solution in this case is

$$x(t) = x_0 e^{-\Gamma(t)} + \int_0^t e^{\Gamma(s)-\Gamma(t)} g(s) dW(s),$$

(3.48)

where

$$\Gamma(t) = \int_0^t \gamma(s) ds.$$

(3.49)

3.6.2 The full linear stochastic equation

This is the equation

$$dx = -\gamma x dt + g x dW.$$

(3.50)

To solve it we can do one of two things. We can change variables to $y = \ln x$, or we can use the more direct approach that we used in Section 2.4. Using the latter method, we rewrite the differential equation as an exponential:

$$x(t + dt) = x + dx = [1 - \gamma dt + g dW]x = e^{-[\gamma + g^2/2]dt + g dW} x,$$

(3.51)

where we have been careful to expand the exponential to second order in dW. We now apply this relation repeatedly to $x(0)$ to obtain the solution. For clarity we choose a finite time-step $\Delta t = t/N$, and we have

$$x(t) = \lim_{N \to \infty} \left(\prod_{n=1}^N e^{-(\gamma + g^2/2)\Delta t + g \Delta W_n} \right) x(0)$$

$$= \lim_{N \to \infty} \exp\left\{ -(\gamma + g^2/2)N\Delta t + g \sum_n \Delta W_n \right\} x(0)$$

$$= \exp\left\{ -(\gamma + g^2/2)t + g \int_0^t dW \right\} x(0)$$

$$= e^{-(\gamma + g^2/2)t + g W(t)} x(0).$$

(3.52)

The random variable $x(t)$ is therefore the exponential of a Gaussian random variable.

We can use the same method to solve the linear stochastic equation when γ and g are functions of time, and we leave the details of this calculation as an exercise. The solution is

$$x(t) = e^{-\Gamma(t) - H(t) + Z(t)} x(0),$$

(3.53)

where

$$\Gamma(t) = \int_0^t \gamma(s)ds, \tag{3.54}$$

$$H(t) = (1/2) \int_0^t g^2(s)ds, \tag{3.55}$$

and $Z(t)$ is the random variable

$$Z(t) = \int_0^t g(s)dW(s). \tag{3.56}$$

3.6.3 Ito stochastic integrals

To summarize the results of Section 3.6.1 above, the stochastic integral

$$I(t) = \int_0^t f(s)dW(s) = \lim_{N \to \infty} \sum_{n=0}^{N-1} f(n\Delta t)\Delta W_n, \tag{3.57}$$

where $f(t)$ is a deterministic function, is a Gaussian random variable with mean zero and variance

$$V[I(t)] = \int_0^t f^2(s)ds. \tag{3.58}$$

We now pause to emphasize an important fact about the stochastic integral in Eq. (3.57). Each term in the discrete sum contains a Wiener increment for a time-step Δt, multiplied by a function evaluated at the *start* of this time-step. This is a direct result of the way we have defined a stochastic differential equation – in fact, solutions to stochastic differential equations, as we have defined them, always involve integrals in which the function in the integral (the integrand) is evaluated at the start of each time-step. This has not been too important so far, but will become important when the integrand is itself a sum of Wiener increments. For example, if $f(t)$ were itself a stochastic process, then it would be a function of all the increments ΔW_j up until time t. So the fact that $f(t)$ is evaluated at the start of the interval means that, for every term in the sum above, ΔW_n is *independent* of all the increments that contribute to $f(n\Delta t)$. Because of this we have

$$\langle f(n\Delta t)\Delta W_n \rangle = \langle f(n\Delta t) \rangle \langle \Delta W_n \rangle = 0, \tag{3.59}$$

and thus $\langle I(t) \rangle = 0$, even if $f(t)$ is a stochastic process. This will be important when we consider multiple stochastic integrals in Section 3.8.3, and modeling multiplicative noise in real systems in Section 5.3.

There are, in fact, other ways to define SDEs so that the resulting stochastic integrals are not defined in terms of the values of the integrand at the start of each interval. They could, for example, be defined in terms of the values of the integrand in the center of each integral, or even at the end of each interval. Because there is more than one way to define a stochastic integral, those in which the integrand is evaluated at the start of each interval are called *Ito* stochastic integrals, and the corresponding SDEs *Ito stochastic equations*. The other versions of stochastic equations are much harder to solve. The main alternative to the Ito stochastic integral is called the *Stratonovich integral*, and as we will see in Section 5.3, this is useful for a specific purpose. Nevertheless, we note now that the alternative ways of defining stochastic integrals, and thus of defining SDEs, all give the same overall class of solutions – the various kinds of SDEs can always be transformed into Ito SDEs.

3.7 Deriving equations for the means and variances

So far we have calculated the means and variances of a stochastic process, x, at some time t, by first solving the SDE for x to obtain the probability density $P(x, t)$. However, there is another method that is sometimes useful if we are only interested in the low moments. If we have the SDE

$$dx = f(x, t)dt + g(x, t)dW \tag{3.60}$$

then we can obtain the differential equation for the mean of x by taking averages on both sides. This gives

$$d\langle x \rangle = \langle dx \rangle = \langle f(x, t) \rangle dt \tag{3.61}$$

because $\langle dW \rangle = 0$. We can rewrite this as

$$\frac{d\langle x \rangle}{dt} = \langle f(x, t) \rangle. \tag{3.62}$$

As an example of using this trick consider the linear SDE

$$dx = -\gamma x dt + \beta x dW. \tag{3.63}$$

We immediately obtain the equation for the mean, being

$$\frac{d\langle x \rangle}{dt} = -\gamma \langle x \rangle. \tag{3.64}$$

Thus

$$\langle x(t) \rangle = e^{-\gamma t} \langle x(0) \rangle. \tag{3.65}$$

To get the equation for the second moment, we first need to calculate the equation for x^2. This is

$$d(x^2) = -2\gamma x^2 dt + 2\beta x^2 dW + \beta^2 x^2 dt. \tag{3.66}$$

By taking the mean on both sides we get the differential equation for $\langle x^2 \rangle$, which is

$$d \langle x^2 \rangle = -(2\gamma - \beta^2) \langle x^2 \rangle dt. \tag{3.67}$$

We can now obtain the differential equation for the variance, by first noting that

$$\frac{dV_x}{dt} = \frac{d \langle x^2 \rangle}{dt} - \frac{d \langle x \rangle^2}{dt} = \frac{d \langle x^2 \rangle}{dt} - 2 \langle x \rangle \frac{d \langle x \rangle}{dt}. \tag{3.68}$$

So using the differential equations that we have derived for the mean and second moment, that for the variance is

$$\frac{dV_x}{dt} = -(2\gamma - \beta^2) \langle x^2 \rangle + 2\gamma \langle x \rangle^2 = -(2\gamma - \beta^2)V_x + \beta^2 \langle x \rangle^2. \tag{3.69}$$

Since we know $\langle x(t) \rangle$ we can solve this using the techniques in Chapter 2.

In general, for a nonlinear stochastic equation, the equation of motion for the mean will contain the second moment(s), and those for the second moments will include the third moments, and so on. The result is an infinite hierarchy of equations, which cannot be solved exactly.

3.8 Multiple variables and multiple noise sources

3.8.1 Stochastic equations with multiple noise sources

Stochastic equations can, of course, be driven by more than one Gaussian noise source. These noise sources can be independent, or mutually correlated, but as we now explain, all correlated Gaussian noise sources can be obtained in a simple way from independent sources. We can solve stochastic differential equations driven by two (or more) independent Wiener noises using the same methods as those described above for a single noise source, along with the additional rule that the product of the increments of different noise sources are zero. As an example, to solve the equation

$$dx = f(x, t)dt + g_1(x, t)dW_1 + g_2(x, t)dW_2, \tag{3.70}$$

the Ito rules are

$$(dW_1)^2 = (dW_2)^2 = dt, \tag{3.71}$$

$$dW_1 dW_2 = 0. \tag{3.72}$$

To obtain two Wiener noise processes that are correlated, all we need to do is to form linear combinations of independent Wiener processes. If we define noise sources dV_1 and dV_2 by

$$\begin{pmatrix} dV_1 \\ dV_2 \end{pmatrix} = M \begin{pmatrix} dW_1 \\ dW_2 \end{pmatrix} = \begin{pmatrix} \sqrt{1-\eta^2} & \eta \\ \eta & \sqrt{1-\eta^2} \end{pmatrix} \begin{pmatrix} dW_1 \\ dW_2 \end{pmatrix}, \tag{3.73}$$

with $-1 \le \eta \le 1$, then dV_1 and dV_2 are correlated even though dW_1 and dW_2 are not. The covariance matrix for dW_1 and dW_2 is $I dt$ (where I is the two-by-two identity matrix), and that for dV_1 and dV_2 is

$$\begin{pmatrix} \langle (dV_1)^2 \rangle & \langle dV_1 dV_2 \rangle \\ \langle dV_1 dV_2 \rangle & \langle (dV_2)^2 \rangle \end{pmatrix} = \left\langle \begin{pmatrix} dV_1 \\ dV_2 \end{pmatrix} (dV_1 , dV_2) \right\rangle$$

$$= \left\langle M \begin{pmatrix} dW_1 \\ dW_2 \end{pmatrix} (dW_1 , dW_2) M^{\mathrm{T}} \right\rangle$$

$$= \begin{pmatrix} 1 & C \\ C & 1 \end{pmatrix} dt, \tag{3.74}$$

where

$$C = 2\eta\sqrt{1-\eta^2}. \tag{3.75}$$

Since the means of dV_1 and dV_2 are zero, and because we have pulled out the factor of dt in Eq. (3.74), C is in fact the correlation coefficient of dV_1 and dV_2.

The set of Ito calculus relations for dV_1 and dV_2 are given by essentially the same calculation:

$$\begin{pmatrix} (dV_1)^2 & dV_1 dV_2 \\ dV_1 dV_2 & (dV_2)^2 \end{pmatrix} = M \begin{pmatrix} dW_1 \\ dW_2 \end{pmatrix} (dW_1 , dW_2) M^{\mathrm{T}}$$

$$= \begin{pmatrix} 1 & C \\ C & 1 \end{pmatrix} dt. \tag{3.76}$$

If we have a stochastic equation driven by two correlated noise sources, such as

$$dx = f(x,t)dt + g_1(x,t)dV_1 + g_2(x,t)dV_2 = f(x,t)dt + \mathbf{g} \cdot \mathbf{dV}, \tag{3.77}$$

with $\mathbf{dV} \equiv (dV_1, dV_2)^{\mathrm{T}}$, then we can always rewrite this in terms of the uncorrelated noises:

$$dx = f(x,t)dt + \mathbf{g} \cdot \mathbf{dV} = f(x,t)dt + \mathbf{g}^{\mathrm{T}} M \mathbf{dW}, \tag{3.78}$$

where we have defined $\mathbf{dW} = (dW_1, dW_2)^{\mathrm{T}}$.

More generally, we can always write N correlated Gaussian noise processes, $\mathbf{dV}$, in terms of N independent Wiener processes, $\mathbf{dW}$. If we want the processes $\mathbf{dV}$ to have the covariance matrix C, so that

$$\mathbf{dV}\mathbf{dV}^{\mathrm{T}} = Cdt, \tag{3.79}$$

then we define

$$d\mathbf{V} = M d\mathbf{W}, \tag{3.80}$$

where M is the square root of C. One can calculate the symmetric matrix M from C by using the definition of a function of a matrix given in Chapter 2: one diagonalizes C and then takes the square root of all the eigenvalues to construct M. (*Note.* Actually M does not have to be symmetric. If M is not symmetric, then $C = MM^{\mathrm{T}}$.)

To summarize, when considering stochastic equations driven by multiple Gaussian noise sources, we only ever need to consider equations driven by independent Wiener processes.

3.8.2 Ito's formula for multiple variables

A general Ito stochastic differential equation that has multiple variables can be written in the vector form

$$d\mathbf{x} = \mathbf{f}(\mathbf{x}, t)dt + G(\mathbf{x}, t)d\mathbf{W}. \tag{3.81}$$

Here $\mathbf{x}$, $\mathbf{f}$, and $d\mathbf{W}$ are the vectors

$$\mathbf{x} = \begin{pmatrix} x_1 \\ x_2 \\ \vdots \\ x_N \end{pmatrix} \quad \mathbf{f} = \begin{pmatrix} f_1(\mathbf{x}, t) \\ f_2(\mathbf{x}, t) \\ \vdots \\ f_N(\mathbf{x}, t) \end{pmatrix} \quad d\mathbf{W} = \begin{pmatrix} dW_1 \\ dW_2 \\ \vdots \\ dW_M \end{pmatrix}, \tag{3.82}$$

where the dW_i are a set of mutually independent noise sources. They satisfy

$$dW_i dW_j = \delta_{ij} dt. \tag{3.83}$$

The symbol G is the $N \times M$ matrix

$$G(\mathbf{x}, t) = \begin{pmatrix} G_{11}(\mathbf{x}, t) & G_{12}(\mathbf{x}, t) & \cdots & G_{1M}(\mathbf{x}, t) \\ G_{21}(\mathbf{x}, t) & G_{22}(\mathbf{x}, t) & \cdots & G_{2M}(\mathbf{x}, t) \\ \vdots & \vdots & \ddots & \\ G_{N1}(\mathbf{x}, t) & G_{N2}(\mathbf{x}, t) & \cdots & G_{NM}(\mathbf{x}, t) \end{pmatrix}. \tag{3.84}$$

To determine Ito's formula for transforming stochastic equations involving multiple variables, all we have to do is use the multi-variable Taylor expansion. Let us say that we wish to transform from a set of variables $\mathbf{x} = (x_1, \ldots, x_N)^{\mathrm{T}}$, to a set of variables $\mathbf{y} = (y_1, \ldots, y_L)^{\mathrm{T}}$, where each of the y_i is a function of some or all of the

x_i, and of time, t. Using the Taylor expansion we have

$$dy_i = \sum_{j=1}^{N} \frac{\partial y_i}{\partial x_j} dx_j + \frac{\partial y_i}{\partial t} dt + \frac{1}{2} \sum_{k=1}^{N} \sum_{j=1}^{N} \frac{\partial^2 y_i}{\partial x_k \partial x_j} dx_k dx_j, \quad i = 1, \ldots, L.$$

(3.85)

This is all one requires to transform variables for any SDE. If we substitute the multivariate SDE given by Eq. (3.81) into the Taylor expansion, and use the Ito rules given in Eq. (3.83) we obtain

$$dy_i = \sum_{j=1}^{N} \frac{\partial y_i}{\partial x_j} \left(f_j(\mathbf{x}, t)dt + \sum_{k} G_{jk}(\mathbf{x}, t)dW_k \right) + \frac{\partial y_i}{\partial t} dt$$

$$+ \frac{1}{2} \sum_{k=1}^{N} \sum_{j=1}^{N} \frac{\partial^2 y_i}{\partial x_k \partial x_j} \left(\sum_{m=1}^{M} G_{jm} G_{km} \right) dt, \quad i = 1, \ldots, L. \quad (3.86)$$

3.8.3 Multiple Ito stochastic integrals

As we have seen in various examples above, the integrals that appear in the solutions to Ito stochastic differential equations for a single variable have the form

$$\int_0^t f(s)dW(s) = \lim_{N \to \infty} \sum_{n=0}^{N-1} f(n\Delta t)\Delta W_n, \quad (3.87)$$

for some function $f(t)$, where $\Delta t = t/N$. In the above summation $f(n\Delta t)$ is the value of the function $f(t)$ at the *start* of the time interval to which the Wiener increment ΔW_n corresponds. As we will see below, this fact becomes important when evaluating multiple stochastic integrals.

When solving Ito equations that have multiple variables, or multiple noise processes, the solutions are in general multiple integrals that involve one or more Wiener processes. It is therefore useful to know how to handle such integrals. For example, we might be faced with the integral

$$I = \int_0^T \int_0^t f(s)dW(s)dt. \quad (3.88)$$

At first sight this may look confusing, but it is easy to evaluate. By discretizing the integral (that is, writing it as a double summation) it becomes clear that we can exchange the order of integration, just as we can for a regular integral. Thus

$$I = \int_0^T \int_t^T ds f(t)dW(t) = \int_0^T (T - t)f(t)dW(t). \quad (3.89)$$

Since this is merely a function of t integrated over the Wiener process, we already know from Section 3.6.1 that it is a Gaussian random variable with mean zero and variance

$$I = \int_0^T [(T - t)f(t)]^2 dt. \tag{3.90}$$

Now, what about double Ito integrals in which both integrals are over Wiener processes? In this case we have two possibilities: the two integrals may involve the *same* Wiener process, or two independent Wiener processes. Note that in both cases we have a double sum of *products* of Wiener increments, so the resulting random variable is no longer Gaussian. In general there is no analytic solution to integrals of this form, but it is not difficult to calculate their expectation values. Before we show how to do this, we note that there is a special case in which a double integral of a single Wiener process can be evaluated exactly. This special case is

$$I = \int_0^T \left[\int_0^t f(s)dW(s) \right] f(t)dW(t). \tag{3.91}$$

To solve this we first define the random variable $Z(t)$ as the value of the inner integral:

$$Z(t) = \int_0^t f(s)dW(s). \tag{3.92}$$

Note that $Z(t)$ is Gaussian with zero mean and variance $V = \int_0^t f^2(s)ds$. We now discretize the double integral:

$$I = \int_0^T Z(t)\,dW(t) = \lim_{\Delta t \to 0} \sum_{n=0}^{N-1} Z_n f_n \Delta W_n, \tag{3.93}$$

where $f_n \equiv f(n\Delta t)$, and rewrite the sum as follows:

$$I = \lim_{\Delta t \to 0} \sum_{n=0}^{N-1} (Z_n + f_n \Delta W_n)^2 - Z_n^2 - (f_{n-1}\Delta W_n)^2$$

$$= \lim_{\Delta t \to 0} \sum_{n=0}^{N-1} \Delta(Z_n^2) - f_n^2(\Delta W_n)^2 = \int_0^T d(Z_n^2(t)) - \int_0^T f^2(t)(dW)^2$$

$$= Z^2(T) - \int_0^T f^2(t)dt. \tag{3.94}$$

This is an exact expression for I, since we can easily obtain an exact expression for the probability density of Z^2.

While we cannot obtain an analytic expression for the integral

$$J = \int_0^T \left[\int_0^T f(s)dW(s) \right] g(t)dW(t) \tag{3.95}$$

(where once again both integrals contain increments of the same Wiener process), we can calculate the expectation value of J. Note that in the definition of J we have fixed the upper limits of both the inner and outer integrals to be T. We will consider replacing the upper limit of the inner integral with t shortly. Discretizing J, and taking the average, we have

$$\langle J \rangle = \lim_{N \to \infty} \sum_{n=0}^{N-1} \sum_{m=1}^{N} f_m g_n \langle \Delta W_m \Delta W_n \rangle$$

$$= \lim_{N \to \infty} \sum_{n=0}^{N-1} f_n g_n \langle (\Delta W_n)^2 \rangle = \int_0^T f(t)g(t)dt. \tag{3.96}$$

We get this result because $\langle \Delta W_m \Delta W_n \rangle$ is Δt if $m = n$, and zero otherwise.

Now let us see what happens if the upper limit of the inner integral is equal to the variable in the outer integral. In this case, the inner integral, being $Z(t) = \int_0^t f(s)dW(s)$, only contains increments of the Wiener process up until time t, and the Wiener increment in the outer integral that multiplies $Z(t)$ is the increment just *after* time t. (It is the increment from t to $t + dt$.) Discretizing the integral we have

$$\langle I \rangle = \lim_{N \to \infty} \sum_{n=1}^{N} \left[\sum_{m=1}^{n-1} f_{m-1} g_{n-1} \langle \Delta W_m \Delta W_n \rangle \right]. \tag{3.97}$$

Thus there are no terms in the double sum that have a product of Wiener increments in the *same* time interval, only those in different time intervals. As a result the expectation value is zero:

$$\langle I \rangle = \left\langle \int_0^T \left[\int_0^t f(s)dW(s) \right] g(t)dW(t) \right\rangle = 0. \tag{3.98}$$

We can encapsulate the above results, Eq. (3.96) and Eq. (3.98), with the rule

$$\langle dW(s)dW(t) \rangle \equiv \delta(t - s)dsdt, \tag{3.99}$$

but this works *only* with the following additional rule: when one of the arguments in the δ-function is equal to the upper or lower limit of the integral, we have

$$\int_L^U \delta(t - L)dt = f(L), \tag{3.100}$$

$$\int_L^U \delta(t - U)dt = 0. \tag{3.101}$$

This means that, for Ito integrals, we count all the area, or "weight", of the δ-function in Eq. (3.99) at the lower limit of an integral, and none at the upper limit.

3.8.4 The multivariate linear equation with additive noise

A multivariate linear system with additive noise is also referred to as a multivariate Ornstein–Uhlenbeck process. When there is only a single noise source this is described by the vector stochastic equation

$$dx = Fxdt + g(t)dW, \tag{3.102}$$

where x and g are vectors, and F is a matrix. In addition, F is a constant, and $g(t)$ is a function of t but not of x. To solve this we can use exactly the same technique as that used above for the single variable Ornstein–Uhlenbeck process, employing what we know about solving linear vector equations from Chapter 2. The solution is

$$\mathbf{x}(t) = e^{Ft}\mathbf{x}(0) + \int_0^t e^{F(t-s)}\mathbf{g}(s)dW(s). \tag{3.103}$$

We can just as easily solve the multivariate Ornstein–Uhlenbeck process when there are multiple independent noises driving the system. In this case the stochastic equation is

$$dx = Fxdt + G(t)\mathbf{dW}, \tag{3.104}$$

where $\mathbf{dW} = (dW_1, dW_2, \ldots, dW_N)$ is a vector of mutually independent Wiener noises dW_i. The solution is obtained in exactly the same way as before, and is

$$\mathbf{x}(t) = e^{Ft}\mathbf{x}(0) + \int_0^t e^{F(t-s)}G(s)\mathbf{dW}(s). \tag{3.105}$$

3.8.5 The full multivariate linear stochastic equation

The multivariate linear stochastic for a single noise source is

$$dx = Fxdt + GxdW, \tag{3.106}$$

and we will take the matrices F and G to be constant. It is not always possible to obtain an analytic solution to this equation. Whether or not an analytic solution exists depends on the *commutator*, or *commutation relations*, between F and G. Many readers will not be familiar with the term "commutation relations", so we now explain what this means.

First, the *commutator* of F and G is defined as

$$[F, G] \equiv FG - GF. \tag{3.107}$$

If $[F, G] = 0$ then $FG = GF$, and F and G are said to *commute*. Two arbitrary matrices F and G in general do not commute. The reason that the commutator of F and G is important, is because, when F and G do not commute, it is no longer true that $e^F e^G = e^{F+G}$. Recall that when we solved the linear stochastic equation for a single variable, we used the fact that $e^a e^b = e^{a+b}$, which is always true when a and b are numbers. This is still true for matrices when they commute, and as a result the solution to Eq. (3.106) when F and G commute is

$$x(t) = e^{(F+G^2/2)t+GW(t)}x(0), \qquad [F, G] = 0. \tag{3.108}$$

To solve the vector linear stochastic equation when F and G do not commute, we need to know how $e^F e^G$ is related to e^{F+G}. This relationship is called the Baker–Campbell–Hausdorff (BCH) formula, and is

$$e^F e^G = e^{F+G+Z_2+Z_3+Z_4+\cdots+Z_\infty}, \tag{3.109}$$

where the additional terms in the sum are "repeated" commutators of F and G. The first two of these terms are

$$Z_2 = (1/2)[F, G], \tag{3.110}$$

$$Z_3 = (1/12)[F, [F, G]] + (1/12)[[F, G], G], \tag{3.111}$$

and higher terms become increasingly more complex [9, 10]. Because of this we cannot obtain a closed-form solution to Eq. (3.106) for every choice of the matrices F and G. However, if the commutator, or some repeated commutator, of F and G is sufficiently simple, then the infinite series in the BCH formula terminates at some point so that it has only a finite number of terms.

We illustrate how to solve Eq. (3.106) for the simplest example in which F and G do not commute. This is the case in which $[F, G]$ commutes with both F and G. That is

$$[F, [F, G]] = [G, [F, G]] = 0. \tag{3.112}$$

In this case Z_3 is zero. Because all higher-order terms involve the commutators of F and G with the terms in Z_3, these vanish too, and the only additional term that remains is $Z_2 = [F, G]/2$. The relationship between $e^F e^G$ and e^{F+G} becomes

$$e^F e^G = e^{F+G+[F,G]/2} = e^{[F,G]/2}e^{F+G}. \tag{3.113}$$

The second relation is true because $F + G$ commutes with $[F, G]$. Also, since $[F, G] = -[G, F]$ we have

$$e^F e^G = e^{F+G-[F,G]/2} = e^{-[F,G]/2} e^{F+G}, \tag{3.114}$$

and putting the above relations together gives

$$e^F e^G = e^{[F,G]} e^G e^F. \tag{3.115}$$

To solve Eq. (3.106) we first write it as

$$\mathbf{x}(t + dt) = e^{\tilde{F}dt + GdW} \mathbf{x}(t) = e^{CdWdt} e^{\tilde{F}dt} e^{GdW} \mathbf{x}(t), \tag{3.116}$$

where we have defined $C \equiv [F, G]$ and $\tilde{F} \equiv F - G^2/2$ to simplify the notation. Here we have used $(dW)^2 = dt$ and the above relations for the matrix exponential. We can now write the solution as

$$\mathbf{x}(t) = \lim_{N \to \infty} \Pi_{n=1}^{N} \left(e^{C \Delta W_n \Delta t} e^{\tilde{F} \Delta t} e^{G \Delta W_n} \right) \mathbf{x}(0), \tag{3.117}$$

where as usual $\Delta t = t/N$. Now we must turn the products of exponentials into exponentials of sums, just like we did to solve the single variable equation. Let us see what to do when we have only two time-steps. In this case we have

$$\mathbf{x}(2\Delta t) = e^{C(\Delta W_1 + \Delta W_2)\Delta t} e^{\tilde{F} \Delta t} e^{G \Delta W_2} e^{\tilde{F} \Delta t} e^{G \Delta W_1} \mathbf{x}(0). \tag{3.118}$$

Note that because C commutes with all the other operators, we can always place all the terms containing C to the left, and combine them in a single exponential. We need to swap the middle two terms, $e^{G \Delta W_2}$ and $e^{\tilde{F} \Delta t}$. If we do this then we have all the terms containing $\tilde{F}$ to the left, which allows us to combine them, and all those containing the ΔWs to the right, also allowing us to combine them. When we perform the swap, we get an additional term $e^{-C \Delta t \Delta W}$ because $[G, \tilde{F}] = [G, F] = -C$. The result is

$$\mathbf{x}(2\Delta t) = e^{C(\Delta W_1 + 2\Delta W_2)\Delta t} e^{-C \Delta W_2 \Delta t} e^{\tilde{F} \Delta t} e^{\tilde{F} \Delta t} e^{G \Delta W_2} e^{G \Delta W_1} \mathbf{x}(0)$$

$$= e^{C(\Delta W_1 + \Delta W_2)} e^{-C \Delta W_2 \Delta t} e^{\tilde{F} 2\Delta t} e^{G(\Delta W_1 + \Delta W_2)} \mathbf{x}(0). \tag{3.119}$$

To obtain $x(3\Delta t)$ we multiply this on the left by $e^{C \Delta W_3 \Delta t} e^{\tilde{F} \Delta t} e^{G \Delta W_3}$. Once again we must swap the two middle terms, bringing the term $e^{\tilde{F} \Delta t}$ to the left. Doing this, which we leave as an exercise, the pattern becomes clear. We can now perform the swap operation N times, and the result is

$$\mathbf{x}(t) = \lim_{N \to \infty} \left(e^{C \Delta t \sum_n \Delta W_n} e^{-C \Delta t \sum_n n \Delta W_n} e^{\tilde{F} t} e^{G \Delta \sum_n W_n} \right) \mathbf{x}(0). \tag{3.120}$$

We can now take the limit as $N \to \infty$. The limits of the various sums are

$$\lim_{N \to \infty} \Delta t \sum_n \Delta W_n = \left(\lim_{\Delta t \to 0} \Delta t \right) \int_0^t dW(s) = 0, \tag{3.121}$$

$$\lim_{N \to \infty} \Delta t \sum_n n \Delta W_n = \lim_{N \to \infty} \sum_n (n \Delta t)(\Delta W_n) = \int_0^t s \, dW(s), \tag{3.122}$$

$$\lim_{N \to \infty} \sum_n \Delta W_n = \int_0^t dW(s) = W(t). \tag{3.123}$$

The solution to Eq. (3.106), when $[F, G]$ commutes with F and G, is thus

$$\mathbf{x}(t) = e^{(F - G^2/2)t} e^{GW(t)} e^{-[F,G]Z(t)} \mathbf{x}(0), \tag{3.124}$$

where we have defined $Z(t) = \int_0^t s \, dW(s)$. The mean of $Z(t)$ is zero, and the variance is $t^3/3$.

Note that in Section 3.4 above we said that we could always set the product $dW\,dt$ to zero. This is true if this product appears in a stochastic differential equation, and this is usually all that matters. However, we see that in implementing the method we have described above for deriving an analytic solution to an SDE we could not drop terms proportional to $dW\,dt$: we had to keep these terms because during the repeated swap operations enough of them where generated to produce a non-zero result.

3.9 Non-anticipating functions

Before we finish with this chapter, it is worth defining the term *non-anticipating function*, also known as an *adapted process*, since it is used from time to time in the literature. Because a stochastic process, $x(t)$, changes randomly in each time-step dt, its value at some future time T is not known at the initial time. Once the future time has been reached, then all the values of the stochastic increments up until that time have been chosen, and the value of $x(T)$ is known. So at the initial time our state of knowledge of $x(T)$ is merely the probability density for $x(T)$, but at time T, $x(T)$ has a definite value that has been picked from this density. To sum this up, if the current time is t, then one knows all the random increments, dW, up until time time t, but none beyond that time.

We say that a function of time, $f(t)$, is *non-anticipating*, or *adapted to the process dW*, if the value of the function at time t depends only on the stochastic increments dW up until that time. Thus if $x(t)$ is the solution to a stochastic equation driven by a stochastic process dW, then $x(t)$ is a non-anticipating function, and any function of $f(x, t)$ is also. When solving stochastic equations, one always deals with non-anticipating functions.

We can go further in this direction. At the initial time, $t = 0$, we have some probability density for $x(T)$, which we obtain by solving the SDE for x. However, at a time, t, where $0 < t < T$, then we know the increments up until time t, and thus the value of $x(t)$. While we still do not know the value of $x(T)$, our probability density for it will now be different – since we know some of the stochastic increments in the interval $[0, T]$, we have more information about the likely values of $X(T)$. Our probability density at time t for $x(T)$ is therefore a *conditional* probability density, conditioned on a knowledge of the increments in the interval $[0, t]$. We will discuss more about how one calculates these conditional probability densities in the next chapter.

Further reading

Our presentation of stochastic differential equations here is intentionally non-rigorous. Our purpose is to convey a clear understanding of SDEs and Ito calculus without getting bogged down by mathematical rigor. We discuss the concepts employed in rigorous treatments in Chapter 10. A rigorous mathematical account of Ito's rule and stochastic differential equations driven by Wiener noise may be found in, for example, *Stochastic Differential Equations: An Introduction with Applications* by Bernt Øksendal [11], *Brownian Motion and Stochastic Calculus* by Karatzas and Shreve [12], and the two-volume set *Diffusions, Markov Processes, and Martingales* by Rogers and Williams [13]. The first of these focusses more on modeling and applications, while the second two contain more theorems and details of interest to mathematicians. The techniques for obtaining analytic solutions to stochastic differential equations presented in this chapter are essentially exhaustive, as far as the author is aware. One further example of a solution to a vector linear stochastic equation containing more complex commutation relations than the one solved in Section 3.8.5 is given in [14], where the method presented in that section was introduced.

Exercises

1. By expanding the exponential to second order, show that $e^{\alpha dt - \beta^2 dt/2 + \beta dW} = 1 + \alpha dt + \beta dW$ to first order in dt.

2. If the discrete differential equation for Δx is

$$\Delta x = cx\Delta t + b\Delta W \qquad (3.125)$$

   and $x(0) = 0$, calculate the probability density for $x(2\Delta t)$.

3. If $z(t) = \int_0^t t'^2 dW(t')$, and $x = az + t$, what is the probability density for x?

4. $x(t) = e^{-b(W(t))^2}$.

  (i) What values can x take?
  (ii) What is the probability density for x?
  (iii) Calculate $\langle x^2 \rangle$ using the probability density for W.
  (iv) Calculate $\langle x^2 \rangle$ using the probability density for x.

5. Calculate the expectation value of $e^{\alpha t + \beta W(t)}$.

6. The random variables x and y are given by

$$x(t) = a + [W(t)]^2,$$
$$y(t) = e^{bW(t)}.$$

  (i) Calculate $\langle x(t) \rangle$.
  (ii) Calculate $\langle y(t) \rangle$.
  (iii) Calculate $\langle x(t)y(t) \rangle$.
  (iv) Are $x(t)$ and $y(t)$ independent?

7. The stochastic process x satisfies the stochastic equation

$$dx = -\gamma x dt + g dW. \tag{3.126}$$

  (i) Calculate the differential equations for $\langle x \rangle$ and $\langle x^2 \rangle$.
  (ii) Calculate the differential equation for $V(x) = \langle x^2 \rangle - \langle x \rangle^2$, and solve it.

8. The stochastic differential equation for x is

$$dx = 3a(x^{1/3} - x)dt + 3\sqrt{a}x^{2/3}dW.$$

  (i) Transform variables to $y = x^{1/3}$, so as to get the stochastic equation for y.
  (ii) Solve the equation for y.
  (iii) Use the solution for $y(t)$ to get $x(t)$, making sure that you write $x(t)$ in terms of $x_0 \equiv x(0)$.

9. Solve the stochastic differential equation

$$dx = -\alpha t^2 x dt + g dW, \tag{3.127}$$

and calculate $\langle x(t) \rangle$ and $V[x(t)]$. Hint. To do this, you use essentially the same method as for the Ornstein–Uhlenbeck equation: first solve the differential equation with $g = 0$, and this tells you how to make a transformation to a new variable, so that the new variable is constant when $g = 0$. The stochastic equation for the new variable is then easy to solve.

10. Linear stochastic equation with time-dependent coefficients: obtain the solution to the equation

$$dx = -f(t)x dt + g(t)x dW. \tag{3.128}$$

11. Solve the linear stochastic equation

$$dx = -\gamma x dt + g x dW, \tag{3.129}$$

by first changing variables to $y = \ln x$. Then calculate the probability density for the solution $x(t)$.

12. Solve the stochastic equations

$$dx = p \, dW \tag{3.130}$$

$$dp = -\gamma p \, dt. \tag{3.131}$$

13. Solve the stochastic equations

$$dx = p dt + \beta \, dV \tag{3.132}$$

$$dp = -\gamma \, dW, \tag{3.133}$$

where dV and dW are two mutually independent Wiener processes. That is, $(dV)^2 = (dW)^2 = dt$ and $dV dW = 0$.

14. Solve the stochastic equation

$$\mathbf{dx} = A\mathbf{x}dt + B\mathbf{x}dW, \tag{3.134}$$

where

$$A = \begin{pmatrix} 0 & \omega \\ -\omega & 0 \end{pmatrix}, \quad B = \begin{pmatrix} \beta & 0 \\ 0 & \beta \end{pmatrix}. \tag{3.135}$$

4

Further properties of stochastic processes

We have seen in the previous chapter how to define a stochastic process using a sequence of Gaussian infinitesimal increments dW, and how to obtain new stochastic processes as the solutions to stochastic differential equations driven by this Gaussian noise. We have seen that a stochastic process is a random variable $x(t)$ at each time t, and we have calculated its probability density, $P(x, t)$, average $\langle x(t) \rangle$ and variance $V[x(t)]$. In this chapter we will discuss and calculate some further properties of a stochastic process, in particular its *sample paths, two-time correlation function*, and power spectral density (or *power spectrum*). We also discuss the fact that Wiener noise is *white noise*.

4.1 Sample paths

A *sample path* of the Wiener process is a particular choice (or *realization*) of each of the increments dW. Since each increment is infinitesimal, we cannot plot a sample path with infinite accuracy, but must choose some time discretization Δt, and plot $W(t)$ at the points $n\Delta t$. Note that in doing so, even though we do not calculate $W(t)$ for the points in-between the values $n\Delta t$, the points we plot do lie precisely on a valid sample path, because we know precisely the probability density for each increment ΔW on the intervals Δt. If we chose Δt small enough, then we cannot tell by eye that the resolution is limited. In Figure 4.1 we plot a sample path of the Wiener process.

If we want to plot a sample path of a more complex stochastic process, defined by an SDE that we cannot solve analytically, then we must use a numerical method to solve the SDE. In this case our solution at each time-step will be approximate, and thus the points we plot on the sample path will also be approximate. We discuss how to solve SDEs numerically in Chapter 6.

So far we have mainly thought of a stochastic process, $x(t)$, as a random variable at each time t. However, it is more accurate to think of it as being described by a

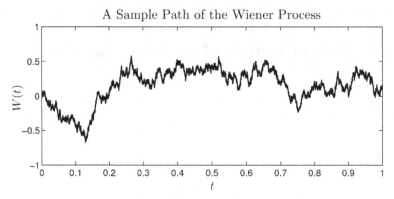

Figure 4.1. A sample path of the Wiener process.

collection of many different possible sample paths. Thus $x(t)$ is actually a random *function*, whose possible values are the different samples paths. If we are only interested in the average value of $x(t)$, then we only need the probability density for x at each time t. If we want to calculate $\langle x(t)x(t+\tau) \rangle$, being the expectation value of the product of x with itself at a later time (which we will do in Section 4.4 below), then we need the joint probability density for x at the two times t and $t+\tau$. How much we need to know about the full probability density for the function $x(t)$ over all the sample paths therefore depends on what we want to calculate.

The sample paths of the Wiener process fluctuate on every time-scale: no matter how small an interval, Δt, we use, because successive Wiener increments ΔW are independent, $W(t)$ will change randomly from one time-step to the next. The result is that a sample path of the Wiener process is *not differentiable*. We can see this if we attempt to calculate the derivative:

$$\frac{dW}{dt} = \lim_{h\to 0} \frac{W(t+h) - W(t)}{h} = \lim_{h\to 0} \frac{W(h)}{h}. \tag{4.1}$$

But we know that $W(h)$ is of order $\sqrt{h}$ (because the standard deviation of $W(h)$ is $\sqrt{h}$), so we have

$$\frac{dW}{dt} = \lim_{h\to 0} \frac{W(h)}{h} \sim \lim_{h\to 0} \frac{1}{\sqrt{h}} \to \infty. \tag{4.2}$$

The fact that the Wiener process is not differentiable is the reason that it is possible to have the relation $dW^2 = dt$ and thus break the usual rules of calculus. If the Wiener process had a derivative then all second-order terms would vanish, and we would not need to do all our calculations in terms of differentials. We could instead write our differential equations in terms of derivatives such as $dx/dt = -\gamma x + g\,dW/dt$, etc. (In fact, it turns out that there is a special case in which we can use the notation dW/dt, and this will be described in Section 4.6.)

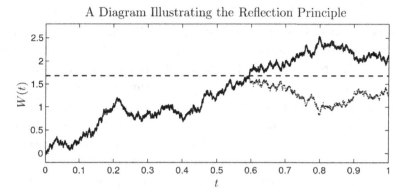

Figure 4.2. Here we show a sample path of the Wiener process that crosses the value a at $t = \tau$. We also show a second sample path that is the same as the first one up until it crosses a, and then is the reflection of the first one about the horizontal line at a (dotted line).

4.2 The reflection principle and the first-passage time

Here we consider the time that it takes the Wiener process to reach a given value, a. (Recall that the Wiener process, by definition, starts with the value zero.) This is called the *first-passage time* to reach the "boundary" a, and we will denote it by T_a. Naturally, since the Wiener process is stochastic, each realization of the process will take a different time to reach the value a. So what we actually want to calculate is the probability density for T_a. This seems at first like a difficult thing to calculate, but it turns out that there is a clever trick (first realized by D. André in 1887) that makes this fairly simple (although the reasoning is a little involved). This trick is called the "reflection principle". We note that our discussion of the reflection principle as a way of calculating first-passage times is included here mainly as a curiosity. In Section 7.7 we will discuss a more generally applicable, and possibly more straightforward way to calculate first-passage times, using Fokker–Planck equations.

Consider a sample path of the Wiener process that crosses the value a at some time τ, as shown in Figure 4.2. We can obtain another sample path from this one by *reflecting* the first path about the horizontal line at height a after the sample path crosses a. This second sample path is the same as the first one up until the crossing time τ, but is different from it after that time, and is shown in Figure 4.2 as a dotted line. Now, the important point about this construction is the following: since the Wiener process is *symmetric* (that is, it is just as likely to go up in any given time-step as to go down) the probability of the first sample path to occur is exactly the same as that of the second. This is the reflection principle. It states that if you reflect the sample path of a symmetric stochastic process from any point onwards, the new path has the same probability of occurring as the first path.

We will now use the reflection principle to calculate the probability density for the first-passage time. Consider a Wiener process that starts with the value 0 at time $t = 0$, and evolves for a time T. During this time it will reach some maximum value that will depend on the sample path that it follows. Call this maximum M_T. Let us now calculate the probability that this maximum M_T is greater than or equal to a. Examining the two paths shown in Figure 4.2, we see that in both cases the maximum value of the Wiener process is greater than or equal to a. However, in only one case (the first path, represented by the solid line) is the final value of the Wiener process greater than a. Thus for every sample path for which the final value of the Wiener process is greater than a, there are *two* paths for which $M_T \geq a$. Therefore we can conclude that

$$\text{Prob}(M_T \geq a) = 2 \times \text{Prob}(W(T) \geq a). \tag{4.3}$$

The probability that $W(T)$ is greater than a is easier to calculate, because we know the probability density for $W(T)$. Thus

$$\text{Prob}(M_T \geq a) = 2 \times \text{Prob}(W(T) \geq a) = \frac{2}{\sqrt{2\pi T}} \int_a^\infty e^{-x^2/(2T)} dx. \tag{4.4}$$

This integral has no analytic solution, but can of course be evaluated numerically.

From the above result we can calculate the probability density for the first-passage time to reach a. We merely have to realize that the time it takes the Wiener process to reach a will be less that or equal to T if and only if the maximum of the Wiener process in time T, M_T is greater than a. Hence we have

$$\text{Prob}(T_a \leq T) = \text{Prob}(M_T \geq a) = \frac{2}{\sqrt{2\pi T}} \int_a^\infty e^{-x^2/(2T)} dx. \tag{4.5}$$

Now the probability that $T_a \leq T$ is simply the distribution function for T_a, $D(T)$, and so the probability density for T_a, $P(T)$, is

$$P(T) = \frac{dD(T)}{dT} = \frac{d}{dT}\left[\sqrt{\frac{2}{\pi T}} \int_a^\infty e^{-x^2/(2T)} dx\right]. \tag{4.6}$$

To evaluate this derivative, we change variables under the integral to $v = (Ta^2)/x^2$ (Note that a is positive, so the integration variable x is always positive, which ensures the mapping from $x \to v$ is well defined). This gives

$$P(T) = \frac{d}{dT}\left[\frac{a}{\sqrt{2\pi}} \int_0^T \frac{e^{-a^2/(2v)}}{v^{3/2}} dv\right], \tag{4.7}$$

and so the fundamental theorem of calculus provides the final result, which is

$$P(T) = \frac{a}{\sqrt{2\pi}} \frac{e^{-a^2/(2T)}}{T^{3/2}}. \tag{4.8}$$

If we calculate the expectation value of T_a (that is, the average time that it will take for the Wiener process to reach the value a) then we find that this is

$$\langle T_a \rangle = \frac{a}{\sqrt{2\pi}} \int_0^\infty \frac{e^{-a^2/(2T)}}{\sqrt{T}} dt = \infty. \tag{4.9}$$

This result for $\langle T_a \rangle$ is reasonable because many paths of the Wiener process go off towards $-\infty$ and never cross a.

4.3 The stationary auto-correlation function, $g(\tau)$

For a stochastic process, it is often useful to know how correlated the values of the process are at two different times. This will tell us how long it takes the process to forget the value it had at some earlier time. We are therefore interested in calculating the correlation coefficient (see Section 1.4)

$$C_{X(t)X(t+\tau)} = \frac{\langle X(t)X(t+\tau)\rangle - \langle X(t)\rangle\langle X(t+\tau)\rangle}{\sqrt{V(X(t))V(X(t+\tau))}} \tag{4.10}$$

for an arbitrary time difference τ. As an illustration we calculate this for the Wiener process. We know already that $V(W(t)) = t$ and thus $V(W(t+\tau)) = t+\tau$. We can calculate the correlation $\langle W(t)W(t+\tau)\rangle$ in the following way:

$$\langle W(t)W(t+\tau)\rangle = \left\langle \int_0^t dW \int_0^{t+\tau} dW \right\rangle = \left\langle \int_0^t dW \left(\int_0^t dW + \int_t^{t+\tau} dW \right) \right\rangle$$

$$= \left\langle \left(\int_0^t dW \right)^2 + \int_0^t dW \int_t^{t+\tau} dW \right\rangle$$

$$= \left\langle \left(\int_0^t dW \right)^2 \right\rangle + \left\langle \int_0^t dW \int_t^{t+\tau} dW \right\rangle \tag{4.11}$$

$$= \langle W(t)^2 \rangle + \left\langle \int_0^t dW \right\rangle \left\langle \int_t^{t+\tau} dW \right\rangle \tag{4.12}$$

$$= \langle W(t)^2 \rangle + \langle W(t)\rangle\langle W(\tau)\rangle = t + 0 = t.$$

And this gives

$$C_{W(t)W(t+\tau)} = \frac{t}{\sqrt{t(t+\tau)}} = \sqrt{\frac{1}{(1+\tau/t)}}. \tag{4.13}$$

In the above derivation, to get from line (4.11) to (4.12) we used the fact that the random variables $A = \int_0^t dW$ and $B = \int_t^{t+\tau} dW$ are *independent*, which implies that their correlation $\langle AB\rangle$ is just the product of their means, $\langle A\rangle\langle B\rangle$. From Eq. (4.13)

we see, as expected, that the Wiener process at time $t + \tau$ is increasingly independent of its value at an earlier time t as τ increases.

The function $g(t, t') = \langle X(t)X(t') \rangle$ is often called the *two-time* correlation function or the *auto*-correlation function ("auto" because it is the correlation of the
process with *itself* at a later time). If the mean of the process $X(t)$ is constant with
time, and the auto-correlation function, $g(t, t + \tau) = \langle X(t)X(t + \tau) \rangle$ is also independent of the time, t, so that it depends only on the *time-difference*, τ, then $X(t)$
is referred to as being "wide-sense" stationary. In this case, the auto-correlation
function depends only on τ, and we write

$$g(\tau) = \langle X(t)X(t + \tau) \rangle. \tag{4.14}$$

The reason that we call a process whose mean and auto-correlation function
are time-independent "wide-sense" stationary, rather than merely "stationary", is
because the latter term is reserved for processes in which the expectation values
of products of the process at *any number* of different times only depends on the
time-differences. Thus while stationary processes are always wide-sense stationary,
wide-sense stationary processes need not be stationary.

The auto-correlation function for a wide-sense stationary process is always
symmetric, so that $g(-\tau) = g(\tau)$. This is easily shown by noting that

$$g(-\tau) = \langle X(t)X(t - \tau) \rangle = \langle X(t - \tau)X(t) \rangle = \langle X(t)X(t + \tau) \rangle = g(\tau). \tag{4.15}$$

4.4 Conditional probability densities

We now describe the general method for calculating two-time correlation functions.
One can always calculate the correlation $\langle X(t')X(t) \rangle$ at two times t' and $t = t' + \tau$,
for some arbitrary process $X(t)$, so long as one has the *joint probability density* that
the value of the process is x at time t and x' and time t'. We will write this density as
$P(x, t; x', t')$. Assume for the sake of definiteness that $t \geq t'$. To calculate this joint
density we first write it in terms of the conditional probability density $P(x, t | x', t')$,
which gives

$$P(x, t; x', t') = P(x, t | x', t')P(x', t'). \tag{4.16}$$

The conditional probability is the probability density for X at time t, given that X
has the value x' at time t'. In fact, we already know how to calculate this, since
it is the same thing that we have been calculating all along in solving stochastic
differential equations: the solution to an SDE for X is the probability density for
X at time t, given that its initial value at $t = 0$ is x_0. To obtain the conditional
probability in Eq. (4.16), all we need to do is solve the SDE for x, but this time
with the initial time being t' rather than 0.

As an example, let us do this for the simplest stochastic equation, $dX = dW$. Solving the SDE means summing all the increments dW from time t' to t, with the initial condition $X(t') = x'$. The solution is

$$X(t) = x' + \int_{t'}^{t} dW = x' + W(t) - W(t'), \tag{4.17}$$

and this has the probability density

$$P(x, t) = \frac{e^{-(x-x')^2/[2(t-t')]}}{\sqrt{2\pi(t-t')}}. \tag{4.18}$$

This is also the conditional probability for X at time t, given that $X = x'$ at time t':

$$P(x, t|x', t') = P(x, t) = \frac{e^{-(x-x')^2/[2(t-t')]}}{\sqrt{2\pi(t-t')}}. \tag{4.19}$$

To calculate the joint probability density we now need to specify the density for X at time t'. If X started with the value 0 at time 0, then at time t' the density for $X(t')$ is just the density for the Wiener process, thus

$$P(x', t') = \frac{e^{-x'^2/[2t']}}{\sqrt{2\pi t'}}. \tag{4.20}$$

Using Eqs. (4.19) and (4.20), the joint density is

$$P(x, t; x', t') = P(x, t|x', t')P(x', t') = \frac{e^{-(x-x')^2/[2(t-t')]-x'^2/[2t']}}{2\pi\sqrt{(t-t')t'}}, \tag{4.21}$$

and the correlation function is therefore

$$\langle X(t')X(t)\rangle = \int_{-\infty}^{\infty}\int_{-\infty}^{\infty} xx'P(x, t; x', t')dxdx' = t'. \tag{4.22}$$

We then obtain the correlation coefficient between $X(t')$ and $X(t)$ by dividing this by the square root of the product of the variances as above.

4.5 The power spectrum

We now return to the Fourier transform introduced in Section 1.8 where we defined the characteristic function. This time we will use a slightly different definition for the Fourier transform: while in Section 1.8 we followed the definition that is standard for characteristic functions, for considering Fourier transforms of signals (functions of time), it is standard to use the following definition. The Fourier transform of the deterministic function $f(t)$ is

$$F(\nu) = \int_{-\infty}^{\infty} f(t)e^{-i2\pi\nu t}\,dt. \tag{4.23}$$

With this definition the inverse transform is

$$f(t) = \int_{-\infty}^{\infty} F(v)e^{i2\pi vt} \, dv. \tag{4.24}$$

Note that this time we have also defined the Fourier transform with a minus sign in the exponential. Since either sign is equally good, which one to use is purely a matter of convention. With this definition for the Fourier transform, the Dirac δ-function is now the Fourier transform of the constant function $f(t) = 1$:

$$\delta(v) = \int_{-\infty}^{\infty} e^{-i2\pi vt} \, dt. \tag{4.25}$$

We now recognize that the inverse Fourier transform expresses a function $f(t)$ as an integral (sum) over the complex exponentials $e^{i2\pi vt}$ with different values of the frequency v. These complex exponentials are complex sinusoids: $e^{i2\pi vt} = \cos(2\pi vt) + i \sin(2\pi vt)$. Thus the inverse Fourier transform expresses a function $f(t)$ as a weighted sum of complex sine waves where the weighting factor is the Fourier transform $F(v)$. This representation for a function of time is very useful.

All the functions of time that we will be interested in here are real, and when discussing Fourier transforms it is common practice to call them *signals*. Since we are concerned with real signals, you might find it a little strange that in representing a function as a sum over sine waves we choose to use complex exponentials rather than real sine and cosine functions. This choice is a very good one, however, because complex exponentials are easier to manipulate than real sines or cosines by themselves.

If $f(t)$ is real, $F(v)$ must be such that all the complex parts of the exponentials cancel when the integral over v is taken. If we write $F(v)$ in terms of its magnitude, $A(v)$, and phase $\phi(v)$, then we have

$$f(t) = \int_{-\infty}^{\infty} A(v)e^{i[2\pi vt + \phi(v)]} \, dv$$

$$= \int_{-\infty}^{\infty} A(v)\cos[2\pi vt + \phi(v)] \, dv + i \int_{-\infty}^{\infty} A(v)\sin[2\pi vt + \phi(v)] \, dv.$$

Since $A(v)$ and $\phi(v)$ are real, the second term above is purely imaginary, and therefore must be zero. So we have

$$f(t) = \int_{-\infty}^{\infty} A(v)\cos[2\pi vt + \phi(v)] \, dv. \tag{4.26}$$

So we see that $f(t)$ is the sum of sine waves of different frequencies v, where the amplitude of the wave at frequency v is given by $A(v)$, and the phase is given by

$\phi(v)$. In terms of $F(v)$, the amplitude and phase are given by

$$A(v) \equiv |F(v)| = \sqrt{F^*(v)F(v)}, \tag{4.27}$$

$$\tan[\phi(v)] = \frac{\text{Im}[F(v)]}{\text{Re}[F(v)]}. \tag{4.28}$$

Here $F^*(v)$ is the complex conjugate of $F(v)$, and

$$\text{Re}[F(v)] = \frac{F(v) + F^*(v)}{2}, \tag{4.29}$$

$$\text{Im}[F(v)] = \frac{F(v) - F^*(v)}{2i}, \tag{4.30}$$

are, respectively, the real and imaginary parts of $F(v)$. The Fourier transform is often referred to as the *spectral decomposition* or *complex spectrum* of a signal $f(t)$, and $A(v)$ as the (real) spectrum.

The square of the value of a signal at time t is referred to as the "instantaneous power" of the signal at that time. This comes from the field of electronics, where the signal is usually a voltage difference between two points. In this case the power dissipated by the current flowing between the two points is proportional to the *square* of the voltage difference between the points, and thus to the square of the signal. This relationship between the signal, $f(t)$, and the instantaneous power is also true for electromagnetic waves (including light and radio waves) as well as sound waves, if the signal is the amplitude of the wave.

4.5.1 Signals with finite energy

If the instantaneous power of the deterministic signal $f(t)$ is defined to be $f^2(t)$, then the total energy of the signal is the integral of $f^2(t)$ over all time:

$$E[f(t)] = \int_{-\infty}^{\infty} f(t)^2 \, dt. \tag{4.31}$$

If the signal has finite duration, then the total energy is finite. Using the definition of the Fourier transform it is not difficult to show that

$$E[f(t)] = \int_{-\infty}^{\infty} f(t)^2 \, dt = \int_{-\infty}^{\infty} |F(v)|^2 \, dv. \tag{4.32}$$

Now, let us say that we pass the signal through a filter that only lets through sinusoids with frequencies between v_1 and v_2. Since the Fourier transform gives the amplitude of each sinusoid that makes up a signal, the signal that emerges from the filter (the filtered signal) would have a Fourier transform of

$$G(v) = \begin{cases} F(v) & v \in [v_1, v_2] \\ 0 & \text{otherwise.} \end{cases} \tag{4.33}$$

The total energy of the filtered signal would then be

$$E[g(t)] = \int_{-\infty}^{\infty} |G(v)|^2 dv = \int_{v_1}^{v_2} |F(v)|^2 dv. \tag{4.34}$$

Thus it makes sense to call $|F(v)|^2$ the spectral energy density of the signal $f(t)$. Once again using the definition of the Fourier transform we find that

$$\begin{aligned}
|F(v)|^2 &= \int_{-\infty}^{\infty} f(t)e^{-i2\pi vt} dt \int_{-\infty}^{\infty} f(t)e^{i2\pi vt} dt \\
&= \int_{-\infty}^{\infty} \left(\int_{-\infty}^{\infty} f(t)f(t-\tau)dt \right) e^{-i2\pi v\tau} d\tau \\
&= \int_{-\infty}^{\infty} \left(\int_{-\infty}^{\infty} f(t)f(t+\tau)dt \right) e^{-i2\pi v\tau} d\tau \\
&= \int_{-\infty}^{\infty} h(\tau)e^{-i2\pi v\tau} d\tau \tag{4.35}
\end{aligned}$$

To obtain the third line we made the transformation $t \to t - \tau$, and in the last line we defined the function

$$h(\tau) = \int_{-\infty}^{\infty} f(t)f(t+\tau)dt. \tag{4.36}$$

Note that $h(t, \tau)$ is a version of an auto-correlation function, but for deterministic signals. This auto-correlation function is the product of the signal with itself at different times, but this time integrated over time rather than averaged over all realizations. We see that the energy spectrum, $|F(v)|^2$, is the Fourier transform of this auto-correlation function.

For a stochastic signal, $x(t)$, the total average energy in the signal is the average value of the instantaneous power, $x^2(t)$, integrated over all time:

$$E[x(t)] = \int_{-\infty}^{\infty} \langle x(t)^2 \rangle \, dt. \tag{4.37}$$

If the mean of $x(t)$ is zero for all t, then of course $\langle x(t)^2 \rangle$ is equal to the variance of $x(t)$.

To calculate the energy spectrum for $x(t)$, we must first define what we mean by the Fourier transform of a stochastic process. A stochastic process, $x(t)$, has many possible sample paths. Recall from Section 4.1 that we can think of $x(t)$ as being described by a probability density over the whole collection of possible sample paths. Each one of these sample paths is a function of time, say $x_\alpha(t)$, where α labels the different possible paths. Thus $x(t)$ is actually a *random function*, whose possible values are the functions $x_\alpha(t)$.

Just as $x(t)$ is a random function, whose values are the sample paths, we define the Fourier transform, $X(v)$, as a random function whose values are the Fourier transforms of each of the sample paths. Thus the possible values of $X(v)$ are the functions

$$X_\alpha(v) = \int_{-\infty}^{\infty} x_\alpha(t)e^{-i2\pi vt}\, dt. \tag{4.38}$$

With this definition of the Fourier transform of a stochastic process, the results above for the energy spectrum are derived in exactly the same way as above for a deterministic process. In particular we have

$$E[x(t)] = \int_{-\infty}^{\infty} \langle|X(v)|^2\rangle\, dv, \tag{4.39}$$

and

$$\langle|F(v)|^2\rangle = \int_{-\infty}^{\infty} \left[\int_{-\infty}^{\infty} g(t, \tau)dt\right] e^{-i2\pi v\tau}d\tau, \tag{4.40}$$

where $g(t, \tau)$ is

$$g(t, \tau) = \langle f(t)f(t + \tau)\rangle. \tag{4.41}$$

4.5.2 Signals with finite power

Many signals we deal with, such as simple sine waves, do not have any specific duration, and we usually think of them as continuing for all time. The energy in such a signal is therefore infinite. This is also true of many stochastic processes – the solution to the Ornstein–Uhlenbeck equation in the long-time limit reaches a stationary solution in which the variance remains constant for the rest of time. For these kinds of signals it is the *average power* of the signal that is finite. For a deterministic process we can define the average power as

$$P = \lim_{T\to\infty} \frac{1}{T}\int_{-T/2}^{T/2} f^2(t)dt, \tag{4.42}$$

and this is finite so long as the signal does not blow up at $\pm\infty$.

We can define the average power of a stochastic signal in the same way as for a deterministic signal, but this time we take the expectation value of the process, so as to average the power both over time and over all realizations (sample paths) of the process. Thus the average power of a stochastic process is

$$P[x(t)] = \lim_{T\to\infty} \frac{1}{T}\int_{-T/2}^{T/2} \langle x^2(t)\rangle dt. \tag{4.43}$$

With the average power defined, it is useful to be able to calculate the *power spectrum*, also called the *power spectral density*, $S(\nu)$. This is defined as the power per unit frequency of a sample path of the process, averaged over all the sample paths. To put this another way, we want a function, $S(\nu)$, that if we were to put a long sample path of the process $x(t)$ through a filter that only lets pass frequencies in the range ν_1 to ν_2, the power of the filtered signal $Y(t)$, averaged over all sample paths, would be

$$P[Y(t)] = \int_{\nu_1}^{\nu_2} S(\nu)d\nu. \tag{4.44}$$

It turns out that the power spectral density of a wide-sense stationary stochastic process $x(t)$ is the Fourier transform of the two-time auto-correlation function. Thus

$$S(\nu) = \int_{-\infty}^{\infty} g(\tau)e^{-i2\pi\nu\tau}d\tau, \qquad g(\tau) = \langle x(t)x(t + \tau)\rangle. \tag{4.45}$$

This result is called the *Wiener–Khinchin theorem*. Proofs of this theorem can be found in [3] and [15].

It is worth noting that if a process is stationary (see Section 4.3 above) we usually do not need to take the average over all sample paths – the average power in a given frequency range, assuming we have long enough sample paths, is usually the same for all sample paths. This is because the average power is defined as the average over *all* time, and averaging a sample path of a wide-sense stationary process over all time is often the same as averaging over all sample paths. If this is true the process is referred to as being *ergodic*. Not all wide-sense stationary processes are ergodic, however, and this can be very difficult to prove one way or the other. If ergodicity cannot be proven for a given physical process, physicists will often assume that it is true anyway.

4.6 White noise

Consider a function $f(t)$ whose integral from $-\infty$ to ∞ is finite. The more sharply peaked f (that is, the smaller the smallest time interval containing the majority of its energy), then the *less* sharply peaked is its Fourier transform. Similarly, the broader a function, then the narrower is its Fourier transform. Now consider a stochastic process $x(t)$. If the auto-correlation function $g(\tau) = \langle x(t)x(t + \tau)\rangle$ drops to zero very quickly as $|\tau|$ increases, then the power spectrum of the process $x(t)$ must be broad, meaning that $x(t)$ contains high frequencies. This is reasonable, since if a process has high frequencies it can vary on short time-scales, and therefore become uncorrelated with itself in a short time.

We have shown above that the derivative of the Wiener process does not exist. However, there is a sense in which the auto-correlation of this derivative exists, and this fact can be useful as a calculational tool. For the sake of argument let us call the derivative of the Wiener function $\xi(t)$. Since the increments of the Wiener process in two consecutive time intervals dt are independent of each other, $\xi(t)$ must be uncorrelated with itself whenever the time separation is greater than zero. Thus we must have $\langle \xi(t)\xi(t+\tau)\rangle = 0$ if $\tau > 0$. In addition, if we try to calculate $\langle \xi(t)\xi(t)\rangle$, we obtain

$$g(0) = \langle \xi(t)\xi(t)\rangle = \lim_{\Delta t \to 0}\left\langle \frac{\Delta W}{\Delta t}\frac{\Delta W}{\Delta t}\right\rangle$$

$$= \lim_{\Delta t \to 0}\frac{\langle(\Delta W)^2\rangle}{(\Delta t)^2} = \lim_{\Delta t \to 0}\frac{1}{\Delta t} \to \infty. \tag{4.46}$$

We recall from Chapter 1 that a function which has this property is the delta-function $\delta(\tau)$, defined by the rule that

$$\int_{-\infty}^{\infty} f(t)\delta(t)dt = f(0) \tag{4.47}$$

for any smooth function $f(t)$. One can think of the delta-function as the limit in which a function with a single peak at $t = 0$ becomes narrower and taller so that the area under the function is always equal to unity. For example,

$$\delta(t) = \lim_{\sigma \to 0}\frac{e^{-t^2/(2\sigma^2)}}{\sqrt{2\pi\sigma^2}}. \tag{4.48}$$

Let us see what happens if we assume that $\xi(t)$ is a noise source with the auto-correlation function $\langle \xi(t)\xi(t+\tau)\rangle = \delta(\tau)$. We now try using this assumption to solve the equation

$$dx = g\,dW. \tag{4.49}$$

Of course we know the solution is just $x(t) = gW(t)$. If $\xi(t)$ exists then we can write the stochastic equation as

$$\frac{dx}{dt} = g\xi(t), \tag{4.50}$$

and the solution is simply

$$x(t) = g\int_0^t \xi(s)ds. \tag{4.51}$$

Now, let us calculate the variance of $x(t)$. This is

$$V(x(t)) = \left\langle g^2 \int_0^t \xi(s)ds \int_0^t \xi(v)dv \right\rangle = g^2 \int_0^t \int_0^t \langle \xi(s)\xi(v)\rangle ds dv$$

$$= g^2 \int_0^t \int_0^t \delta(s-v)ds dv = g^2 \int_0^t dv = g^2 t, \tag{4.52}$$

which is the correct answer. Also, we can calculate the two-time auto-correlation function of $x(t)$. This is

$$\langle x(t)x(t+\tau)\rangle = \left\langle g^2 \int_0^t \xi(s)ds \int_0^{t+\tau} \xi(v)dv \right\rangle = g^2 \int_0^t \int_0^{t+\tau} \delta(s-v)ds dv$$

$$= g^2 \int_0^t dv = g^2 t, \tag{4.53}$$

which is also correct. So we have been able to obtain the correct solution to the stochastic differential equation by assuming that $\xi(t) \equiv dW(t)/dt$ exists and has a delta auto-correlation function. This technique will work for any SDE that has *purely additive noise*, but it does not work for SDEs in which the noise dW multiplies a function of any of the variables. Thus one cannot use it to solve the equation $dx = xdW$, for example. Note also that in using the relation $\langle \xi(t)\xi(t + \tau)\rangle = \delta(\tau)$, we must use the additional definition discussed in Section 3.8.3: if the peak of the δ-function is located at one of the limits of an integral, we must take all its area to be at the lower limit, and none at the upper limit.

Now, since the power spectrum of a process is the Fourier transform of the auto-correlation function, this means that the spectrum of ξ is

$$S(v) = \int_{-\infty}^\infty \delta(t)e^{-i2\pi vt}dt = 1. \tag{4.54}$$

This spectrum is the same for all values of the frequency v. This means that $\xi(t)$ contains arbitrarily high frequencies, and thus infinitely rapid fluctuations. It also means that the power in $\xi(t)$ is infinite. Both of these are further reflections of the fact that $\xi(t)$ is an idealization, and cannot be truly realized by any real noise source. Because the spectrum contains equal amounts of all frequencies, it is referred to as "white" noise.

The reason that we can assume that $\xi(t)$ exists, and use this to solve stochastic equations that contain additive noise, is that a differential equation is a filter that does not let through infinitely high frequencies: even though $\xi(t)$ has infinitely high frequencies, the solution to a differential equation driven by $\xi(t)$ does not. So long as the bandwidth of the dynamics of the system (that is, the frequencies the differential equation "lets through") are small compared to the bandwidth of frequencies of a real noise source that drives it, white noise will serve as a good approximation to the real noise. This is why we can use the Wiener process to model noise in real systems. To derive these results, we would need to discuss linear-system theory, which is beyond our scope here. The question of why we cannot use $\xi(t)$ (at least as defined above) to model real noise that is not additive is more subtle, and will be discussed in section 5.3.

Further reading

Further details regarding δ-functions, Fourier transforms, filters, and power spectra are given in the excellent and concise text *Linear Systems* by Sze Tan [3]. Fourier transforms and filters in deterministic systems may also be found in, for example, *Signals and Systems* by Alan Oppenheim and Alan Willsky [4]. Further details regarding the power spectral density and its applications are given in *Principles of Random Signal Analysis and Low Noise Design: The Power Spectral Density and its Applications*, by Roy Howard [15]. Techniques for detecting signals in noise are discussed extensively in *Detection of Signals in Noise*, by Robert McDonough and Anthony Whalen [16].

Exercises

1. Show that

    (i) If $f(t)$ is real then the Fourier transform satisfies $F(v) = F(-v)^*$.
    (ii) If $F(v)$ is the Fourier transform of $f(t)$, then the Fourier transform of $f(at)$ is $F(v/a)/|a|$.
    (iii) If $F(v)$ is the Fourier transform of $f(t)$, and assuming that $f(t)$ is zero at $t = \pm\infty$, the Fourier transform of $d^n f/dt^n$ is $(i2\pi v)^n F(v)$.

2. Show that

$$\int_{-\infty}^{\infty} f(t)^2 dt = \int_{-\infty}^{\infty} |F(v)|^2 dv. \tag{4.55}$$

    Hint: use the fact that $\delta(t - \tau) = \int_{-\infty}^{\infty} e^{-i2\pi v(t-\tau)} dv$.
3. Calculate the average power of the signal $f(t) = A\sin(\omega t)$.
4. Calculate the auto-correlation function of

$$x(t) = \cos(\omega t)x_0 + g \int_0^t \cos[\omega(t - s)]dW(s), \tag{4.56}$$

    where x_0 is a random variable, independent of the Wiener process, with mean zero and variance V_0.
5. Calculate the auto-correlation function of

$$x(t) = e^{-\beta W(t)} W(t). \tag{4.57}$$

6. (i) Calculate the auto-correlation function of the solution to

$$dx = -\gamma x dt + g dW. \tag{4.58}$$

   (ii) Show that in the limit as $t \to \infty$ the auto-correlation function is independent of the absolute time t. In this limit the process therefore reaches a "steady-state" in which it is wide-sense stationary.

   (iii) In this limit calculate the power spectral density of the process.

7. (i) Calculate the auto-correlation of the solution to

$$dx = -\frac{\beta^2}{2} x dt + \beta x dW. \tag{4.59}$$

   (ii) Show that in the limit as $t \to \infty$ the auto-correlation function is independent of the absolute time t. In this limit the process therefore reaches a "steady-state" in which it is wide-sense stationary.

   (iii) In this limit calculate the power spectral density of the process.

8. Calculate the auto-correlation function of x in the long-time limit, where

$$dx = (\omega y - \gamma x)dt, \tag{4.60}$$

$$dy = (-\omega x - \gamma y)dt + g dW. \tag{4.61}$$

5

Some applications of Gaussian noise

5.1 Physics: Brownian motion

In 1827, a scientist named Robert Brown used a microscope to observe the motion of tiny pollen grains suspended in water. These tiny grains jiggle about with a small but rapid and erratic motion, and this motion has since become known as *Brownian motion* in recognition of Brown's pioneering work. Einstein was the first to provide a theoretical treatment of this motion (in 1905), and in 1908 the French physicist Paul Langevin provided an alternative approach to the problem. The approach we will take here is very similar to Langevin's, although a little more sophisticated, since the subject of Ito calculus was not developed until the 1940s.

The analysis is motivated by the realization that the erratic motion of the pollen grains comes from the fact that the liquid is composed of lumps of matter (molecules), and that these are bumping around randomly and colliding with the pollen grain. Because the motion of the molecules is random, the net force on the pollen grain fluctuates in both size and direction depending on how many molecules hit it at any instant, and whether there are more impacts on one side of it or another.

To describe Brownian motion we will assume that there is a rapidly fluctuating force on the molecule, and that the fluctuations of this force are effectively white noise. (The definition of white noise is given in Section 4.6.) Also, we know from fluid dynamics that the pollen grain will experience friction from the liquid. This friction force is proportional to the negative of the momentum of the grain, so that

$$F_{\text{friction}} = -\gamma p = -\gamma m v, \tag{5.1}$$

where m is the mass of the grain and v is its velocity. The constant of proportionality, γ, is usually referred to as the damping rate, and is given by $\gamma = 6\pi \eta a / m$. Here a is the diameter of the pollen grain (assumed spherical), and η is the viscosity of the liquid.

We will only treat the motion in one dimension, since we lose nothing important by this restriction, and the extension to three dimensions is simple. The equation of motion for the position of the pollen grain is therefore

$$m\frac{d^2x}{dt^2} = F_{\text{friction}} + F_{\text{fluct}} = -\gamma p + g\xi(t). \tag{5.2}$$

Here $\xi(t)$ is the white noise process discussed in Chapter 4, with correlation function

$$\langle \xi(t)\xi(t+\tau)\rangle = \delta(\tau), \tag{5.3}$$

and g is a constant giving the overall magnitude of the fluctuating force. We will determine the value of g shortly. We can write the equation of motion in vector form, as described in Chapter 2, and this is

$$\frac{d}{dt}\begin{pmatrix} x \\ p \end{pmatrix} = \begin{pmatrix} 0 & 1/m \\ 0 & -\gamma \end{pmatrix}\begin{pmatrix} x \\ p \end{pmatrix} + \begin{pmatrix} 0 \\ g\xi(t) \end{pmatrix}. \tag{5.4}$$

Finally we write the equations in differential form, and this gives

$$\begin{pmatrix} dx \\ dp \end{pmatrix} = \begin{pmatrix} 0 & 1/m \\ 0 & -\gamma \end{pmatrix}\begin{pmatrix} x \\ p \end{pmatrix} dt + \begin{pmatrix} 0 \\ g \end{pmatrix} dW, \tag{5.5}$$

where dW is our familiar Wiener process.

Solving this set of equations is quite simple, because while dx depends on p, dp does not depend on x. We first solve the equation for p, being the Ornstein–Uhlenbeck process

$$dp = -\gamma p\, dt + g\, dW. \tag{5.6}$$

Using the techniques in Chapter 3, the solution is

$$p(t) = e^{-\gamma t}p(0) + g\int_0^t e^{-\gamma(t-s)}dW(s). \tag{5.7}$$

The variance of $p(t)$ is thus

$$V(p(t)) = g^2\int_0^t e^{-2\gamma(t-s)}ds = \frac{g^2}{2\gamma}\left(1 - e^{-2\gamma t}\right). \tag{5.8}$$

Now we know from statistical mechanics that the average kinetic energy, $\langle E\rangle = \langle p^2/(2m)\rangle$, of the pollen grain in the steady-state is equal to $k_B T/2$, where k_B is Boltzmann's constant and T is the temperature of the liquid [17]. Thus in order for our model to agree with what we already know about the physics of a small particle in a fluid, the solution must have a steady-state, and must give the correct value for the steady-state average energy. The first of these means that the probability density for $p(t)$ should become independent of time as $t \to \infty$. Examining the solution for $V(p(t))$ above, we see that

$$\lim_{t\to\infty} V(p(t)) = \frac{g^2}{2\gamma} \equiv V(p)_{\text{ss}}, \tag{5.9}$$

and thus the variance does reach a steady-state. Now, since the mean value of $p(t)$ tends to zero as $t \to \infty$, in the steady-state the mean is zero and thus $\langle p^2 \rangle = V(p(t))$. For consistency with statistical mechanics we therefore require that

$$\langle E \rangle = \langle p^2/(2m) \rangle = \frac{V(p)}{2m} = \frac{g^2}{4\gamma m} = \frac{g^2}{24\pi \eta d}, \tag{5.10}$$

and this tells us that the strength of the noise must be

$$g = \sqrt{12\pi \eta d k T}. \tag{5.11}$$

We now want to calculate a quantity that we can actually measure in an experiment. To do this we examine the solution for the position of the pollen grain, $x(t)$. To obtain this all we have to do is integrate $p(t)/m$ with respect to time, and this gives

$$x(t) = \int_0^t \frac{p(s)}{m} ds = \frac{p(0)}{m} \int_0^t e^{-\gamma s} ds + \frac{g}{m} \int_0^t \left[\int_0^s e^{-\gamma(s-s')} dW(s') \right] ds$$

$$= \frac{p(0)}{m} \int_0^t e^{-\gamma s} ds + \frac{g}{m} \int_0^t e^{\gamma s'} \left[\int_{s'}^t e^{-\gamma s} ds \right] dW(s')$$

$$= \frac{1}{m\gamma} (1 - e^{-\gamma t}) p(0) + \frac{g}{m\gamma} \int_0^t (1 - e^{-\gamma(t-s')}) dW(s')$$

$$= \frac{1}{m\gamma} (1 - e^{-\gamma t}) p(0) + \frac{g}{m\gamma} \int_0^t (1 - e^{-\gamma s}) dW(s). \tag{5.12}$$

In the above calculation we switched the order of integration in the double integral. Recall from Section 3.8.3 that this can be done using the usual rules of calculus, a fact which is simple to show by discretizing the integral.

The above expression, Eq. (5.12), is the complete solution for $x(t)$. We see from this that the probability for $x(t)$ is a Gaussian. We can now easily calculate the mean and variance, which are

$$\langle x(t) \rangle = \left\langle \frac{1}{m\gamma} (1 - e^{-\gamma t}) p(0) \right\rangle + \left\langle \frac{g}{m\gamma} \int_0^t (1 - e^{-\gamma s}) dW(s) \right\rangle$$

$$= \frac{1}{m\gamma} (1 - e^{-\gamma t}) p(0), \tag{5.13}$$

$$V(x(t)) = V\left(\frac{1}{m\gamma} (1 - e^{-\gamma t}) p(0) \right) + V\left(\frac{g}{m\gamma} \int_0^t (1 - e^{-\gamma s}) dW(s) \right)$$

$$= \frac{g^2}{(m\gamma)^2} \int_0^t (1 - e^{-\gamma s})^2 ds$$

$$= \frac{g^2 t}{(m\gamma)^2} + \frac{g^2}{2m^2\gamma^3} \left[4e^{-\gamma t} - e^{-2\gamma t} - 3 \right]. \tag{5.14}$$

Langevin realized that the damping (or decay) rate γ was very fast, much faster than the time-resolution with which the particle could be observed in a real experiment (at least back in 1908). So this means that $\gamma t \gg 1$. In this case the variance of $x(t)$ simplifies to the approximate expression

$$V(x(t)) \approx \frac{g^2 t}{(m\gamma)^2} - \frac{3g^2}{2m^2\gamma^3} = \frac{g^2}{(m\gamma)^2}\left(t - \frac{3}{2\gamma}\right) \approx \frac{g^2 t}{(m\gamma)^2} = \left(\frac{kT}{3\pi\eta d}\right)t.$$

(5.15)

Thus, for times at which $t\gamma \gg 1$, the variance of the position of the particle will be proportional to time. This is exactly the same behavior as Wiener noise, and because of this Wiener noise is often referred to as Brownian motion. To test if our model for Brownian motion is accurate, one can perform an experiment in which the motion of a pollen grain is observed over time. If we record the amount by which the position changes in a fixed time interval T, and average the square of this over many repetitions of the experiment, then we can calculate the variance $V(x(T))$. This experiment was performed in 1910 by Smoluchowski and others, confirming the scaling of the variance with time. Since that time many more experiments have been performed, and deviations from Wiener noise have been observed at very short time-scales [18].

So how is it that $x(t)$ can act just like the Wiener process? For this to be the case, it needs to be true that

$$dx \approx \alpha dW, \quad \text{or equivalently} \quad \frac{dx}{dt} \approx \xi(t), \quad\quad (5.16)$$

for some constant α. Since $dx/dt = p/m$, this means that $p(t)$ is effectively acting like $\xi(t)$. The reason for this is that the auto-correlation function of $p(t)$ decays at the rate γ (see Exercise 4 in Chapter 3), so that if γ is large then the auto-correlation function is sharply peaked, and approximates the auto-correlation function of $\xi(t)$, which is a delta-function. Of course, when we say that γ is large, we actually mean that it is large *compared* to something else. In this case the something else is the time resolution (or time-scale), δt, over which the process is observed. By *time-scale* we mean the time that elapses between observations of the process. In this case the process is $x(t)$. If we know the value of $x(t)$ at time $t = 0$, and then again at a later time $t + \delta t$, then we cannot distinguish it from the Wiener process so long as $\delta t \gg 1/\gamma$. So when we say that γ is large enough that we cannot distinguish $p(t)$ from $\xi(t)$, we mean that γ is much larger than $1/\delta t$, where δt is the time resolution of our knowledge of $p(t)$ and $x(t)$.

5.2 Finance: option pricing

The prices of the stocks (alternatively, shares) of companies exhibit unpredictable fluctuations, and are therefore good candidates for modeling by stochastic processes. Because of this, stochastic processes have important applications in finance, in particular in the pricing of financial derivatives. A financial derivative is something whose value "derives" in some way from the price of a basic asset, such as a share in a company. An *option* is one such financial derivative. The simplest kind of option, termed a "European" option, is a contract that gives the bearer the right to buy (or to sell) a number of shares at a fixed price (called the "strike price") at a fixed date in the future. The date in question is called the "maturity time" of the option. Let us assume for the sake of definiteness that the option is an option to buy, called a "call option" (an option to sell is called a "put" option). Now, clearly, if at the maturity time the share price is higher than the strike price, the bearer will exercise his or her option to buy, then immediately sell the shares so obtained at their current value, and pocket the difference! In that case the value of the option at that time is the precisely the amount of money the bearer makes by this transaction. If, on the other hand, at maturity the share price is *lower* than the strike price, the bearer will not exercise the option and the value of the option is zero.

We see that the value of an option at its maturity depends upon the price of the corresponding share (referred to in financial jargon as the "underlying share") at the time of maturity. The problem of option pricing is to calculate the value of the option well *before* maturity, when the contract is first agreed upon. Clearly the option must be worth something, because once the bearer has the option, they stand to profit from it with some probability, and have nothing to lose by it. But how much should they pay for it? It was Fischer Black and Myron Scholes who first worked out how to calculate the value of an option (in 1973), and it is their analysis that we will describe here. (For this work Merton and Scholes were awarded the Bank of Sweden Prize in Memory of Alfred Nobel, often referred to as the "Nobel prize in economics".)

5.2.1 Some preliminary concepts

The interest rate

If we put money in a bank account, then the bank pays us interest on a regular basis, and as a result the money accumulates. Each interest payment is a specified fraction of the money in the account. In finance, one always assumes that any money is stored in a bank account, because it makes no sense not to take advantage of interest payments. If the amount of interest that is paid is k times the present amount, and this interest is paid once each time interval Δt, then the discrete differential equation for the amount of money in the bank account is

$$\Delta M = kM\Delta t. \tag{5.17}$$

As we saw in Chapter 2, the solution to this equation is

$$M(N\Delta t) = \prod_{n=0}^{N-1}(1+k)M(0) = (1+k)^N M(0). \tag{5.18}$$

If we make the accumulation of interest continuous, then we have the differential equation

$$dM = rM dt, \tag{5.19}$$

where r is called the interest rate. In this case our money now evolves as

$$M(t) = e^{rt}M(0). \tag{5.20}$$

In finance one usually assumes that all money increases exponentially like this at a rate r, where r is called the *risk-free* interest rate. It is risk free because the chance of the bank folding and not giving you your money back is assumed to be so small that it can be ignored.

Arbitrage

A market is inconsistent if a single commodity has two different prices, or two effectively equivalent commodities have different prices (we will give an example of two "equivalent" commodities below). Because of this inconsistency one could buy the commodity at the cheaper price, and sell it at the higher price, making a risk-free profit. Making such a profit out of market inconsistencies is called *arbitrage*. When traders buy at the lower price and sell at the higher price in order to make a profit, then this raises the demand for the lower-priced items, which in turn raises their price. Similarly, selling the higher-priced items increases their supply and lowers their price. The result is that the two prices converge together eliminating the inconsistency. Traders who watch the market and take advantage of arbitrage opportunities, called arbitrageurs, thus perform the function of keeping the market consistent. As a result, it is usual in finance to assume that markets are consistent, with any inconsistencies being eliminated quickly as they arise by arbitrageurs.

The assumption that a market is consistent has important consequences for two commodities that are not identical, but which are equivalent, and should therefore have the same price. As an example of this we now consider the concept of a *forward contract*. A forward contract is an agreement to buy a certain item at a specified price, F, at a specified future date. We will call the present date $t = 0$, and the date of the forward contract T. The question is, given that we know that the current price of the item is P dollars, what should the agreed price, F, in the forward contract be? Answering this question is simplest if we assume that you currently own the item. In this case you can sell it for P dollars now, and write a forward contract to buy the same item for F at the future time T. If you put your P dollars in the bank, at the future time they will be worth $P' = e^{rT}P$. When you buy back the item at time T, you will pay F dollars. Thus at time T you will own the item, and have made a risk-free profit of

$$\text{money for nothing} = P' - F = e^{rT}P - F. \tag{5.21}$$

This is positive if $F < e^{rT}P$. We can alternatively cast this ability to lock in a risk-free profit as an arbitrage situation taking advantage of a market inconsistency. Here there are two things that are equivalent both at the present time, and at the future time T. The first is P dollars of money at the present time, and a forward contract to buy an item for a "forward price" F dollars at a future time T. The second is the item whose present price is P dollars. Both of these are equivalent to having the item now, and having the item at the future time. Thus, if these two things do not have the same price at the future time (being the item plus zero dollars) then there is an arbitrage opportunity.

If alternatively the forward price F is greater than $e^{rT}P$, then one can make money from the arbitrage opportunity by being the person who is on the selling end of the forward contract. In this case we borrow P dollars, use it to buy the item, and then write a forward contract to sell the item for F dollars at time T. Assuming we can borrow money at the risk-free interest rate, then at time T we must pay back $e^{rT}P$ dollars, and we receive F dollars from the sale of the item. This time the profit we make is $F - e^{rT}P$.

So if we assume that there are no arbitrage opportunities, this tells us that the forward price one should set for a commodity on a forward contract is the present price multiplied by e^{rT}. That is, $F = e^{rT}P$. If the forward contract is made out for a different forward price, G, then the buyer of the contract (the person who has agreed to buy the commodity) can make $M = Pe^{rT} - G$ dollars at time T. We can make this second situation consistent with no-arbitrage by making the buyer *pay* $e^{-rT}M = P - e^{-rT}G$ dollars for the contract. So this means that a forward contract that is *not* written with a forward price $F = e^{rT}P$ is itself worth something. It has a value itself – let us call this $\mathcal{F}$ – and so must be bought for this price by the buyer of the contract. Note also that once the forward contract has been written with a certain forward price, G, as time goes buy the current price of the commodity will usually change. This means that a forward contract that was initially worth nothing itself, so that $\mathcal{F} = 0$, will be worth something (which could be negative) as time goes by. Because of this, forward contracts are bought and sold up until the date T. Forward contracts that are bought and sold like this are called "futures".

To find the price of an option (for example, a contract that gives the bearer an option to buy or sell a fixed amount of shares at given price at a future date) one similarly uses a "no-arbitrage" argument, and this is the subject of the next section.

5.2.2 Deriving the Black–Scholes equation

To begin with we need to define the term "portfolio", with which most people are probably already familiar: a portfolio is a collection of assets owned by someone, and which for whatever reason we group together as single unit. A portfolio may consist of a variety of shares in various companies, or it could be a collection of more general assets including shares, money in bank accounts, real estate or even contracts such as options.

Now, let us denote the price of a single share of a company by $S(t)$. We have written this as a function of time, since the price of the share will vary as time goes by. We know from experience that the prices of shares fluctuate in an erratic and apparently random fashion. Thus it seems reasonable that we should describe the price of a share as a stochastic process. The question is, which stochastic process? It seems reasonable that the amount by which a share is likely to fluctuate randomly in a given time interval is proportional to its current value. That is, that a share worth 1 dollar would be as likely to move by 10 cents over a period of a week, as a share worth 10 dollars would be to move by 1 dollar. While this may not be exactly true, it is very reasonable because, after all, money is an arbitrary measure of value, and thus the *absolute* value of a share is not a particularly meaningful quantity. We will therefore assume that the random part of the change in S, dS, in time interval dt is given by

$$dS_{\text{random}} = \sigma S dW, \qquad (5.22)$$

where dW is our familiar Wiener process, and σ is a constant. We also know from looking at history that share prices, on average, grow exponentially with time, just like money in a bank account. We will therefore choose the deterministic part of the increment in the share to be one that generates exponential growth. Thus

$$dS = \mu S dt + \sigma S dW = (\mu dt + \sigma dW)S. \qquad (5.23)$$

Here μ is referred to as the *expected rate of return* of the share, and σ is called the *volatility*. The volatility determines how large the random fluctuations are. This is the model that Black and Scholes used to describe the time evolution of a share. Note that it is the time-independent multiplicative stochastic equation for which we found the solution in Chapter 3. The advantage of this model is that it is simple, but we now know, from detailed statistical analyses of historical share prices, that it is not quite right. This is mainly because the assumption that the random fluctuations are Gaussian is not quite right, and there is considerable research devoted to devising more accurate models of the stock market. However, it is useful because it is sufficiently simple that we can derive fairly simple expressions for the prices of some kinds of options, and it is sufficiently realistic that it is still used by traders as a practical tool to calculate option prices.

To work out the price of a call option, C, we are going to construct two equivalent portfolios, one that includes the option, and another whose time dependence we know. From the assumption of no arbitrage, we will then be able to determine the value of the option. To make the first portfolio, we will need to be able to own a *negative* amount of an asset. So what does this mean? This means that when the value of the asset goes up by an amount x, the value of our portfolio goes *down* by the same amount. To make a portfolio like this we do the following. We borrow the asset off somebody, with an agreement to give it back at a time T later, and then immediately sell it (this is called short-selling). If its current price is P, then we now have P dollars. Our portfolio consists of a debt of the asset to the person we borrowed it off, and P dollars in the bank, being the current price of the asset. Now let us see what the value of our portfolio is at time T if the price of the asset has gone up. To be able to give the asset back to the person we borrowed it off, we must buy the asset back. If the new price of the asset is $P + x$, then we have to spend $P + x$ dollars. Thus the money we have after we buy back the asset is *reduced* by x dollars. What is the total value of our portfolio at time T? The money we have has gone up by the interest we earned, which is $(e^{rT} - 1)P$, and has gone *down* by x dollars. Thus, the portfolio behaves just as if we have positive P dollars in the bank, and a *negative* amount of the asset.

To work out the price of an option we make a portfolio out of the option and a carefully chosen amount of the asset, and show that this portfolio has no risk. Since it has no risk, it must be equivalent to having money in the bank. To construct this special portfolio, consider an option that gives the bearer the right to buy or sell an asset, S, whose price is S. We say that the option is "written" on the asset, and that S is the "underlying asset", or the "underlier" for the option. We now make the assumption that the value of the option, C, (whatever it is) will depend only on the value of the underlying asset, S, and time. Thus we can write $C = C(S, t)$. We will also assume that the value of the option is a smooth function of the asset price and of time, so that the derivatives $\partial C/\partial S$, $\partial^2 C/\partial S^2$, and $\partial C/\partial t$ exist. (We will find that this is true.) Now we construct a portfolio consisting of the option and an amount of the underlying asset S. The value of the portfolio is

$$\Pi = C + \alpha S, \tag{5.24}$$

where α is the amount of the asset. We want to work out the change in the value of the portfolio when we increment time by dt. Since we know the stochastic differential equation for S, we can work out the stochastic differential equation for

Π. We have

$$d\Pi = dC + \alpha\, dS$$

$$= C(S + dS, t + dt) - C(S, t) + \alpha\, dS$$

$$= C(S, t) + \frac{\partial C}{\partial S} dS + \frac{1}{2}\frac{\partial^2 C}{\partial S^2}(dS)^2 + \frac{\partial C}{\partial t} dt - C(S, t) + \alpha\, dS$$

$$= \left[\frac{\partial C}{\partial S} + \alpha\right] dS + \frac{1}{2}\frac{\partial^2 C}{\partial S^2}(dS^2) + \frac{\partial C}{\partial t} dt$$

$$= \left[\frac{\partial C}{\partial S} + \alpha\right](\mu S\, dt + \sigma S\, dW) + \frac{1}{2}\frac{\partial^2 C}{\partial S^2}(S^2\sigma^2 dt) + \frac{\partial C}{\partial t} dt$$

$$= \left[\mu S\left(\frac{\partial C}{\partial S} + \alpha\right) + \frac{\partial C}{\partial t} + \frac{1}{2}\sigma^2 S^2\frac{\partial^2 C}{\partial S^2}\right] dt + \sigma S\left[\frac{\partial C}{\partial S} + \alpha\right] dW.$$

$$(5.25)$$

Of course, we cannot solve this stochastic equation, since we don't (yet) know what $\partial C/\partial S$, $\partial^2 C/\partial S^2$ and $\partial C/\partial t$ are. However, we observe from this equation that if we chose the amount of the asset to be

$$\alpha = -\frac{\partial C}{\partial S}, \qquad (5.26)$$

then the equation for dS becomes completely deterministic! That is, if we choose the right amount of the asset, any stochastic change in the price of the option is exactly offset by the change in the price of the asset. In this case the value of the portfolio Π has no random component, and therefore is just like having Π dollars in the bank. Since this is equivalent to money, in order to avoid arbitrage opportunities, the change in the value of this portfolio in the time interval dt must be the same as that of money. Now recall that the "equation of motion" for money M is

$$dM = rM\, dt, \qquad (5.27)$$

where M is the "risk-free" interest rate. We can therefore conclude that when we set $\alpha = -\partial C/\partial S$, then $d\Pi$ should also obey the equation

$$d\Pi = r\Pi\, dt. \qquad (5.28)$$

Equating the right-hand sides of Eq. (5.28) and Eq. (5.25), we have

$$r\Pi\, dt = \left[\frac{\partial C}{\partial t} + \frac{1}{2}\frac{\partial^2 C}{\partial S^2}S^2\sigma^2\right] dt. \qquad (5.29)$$

Dividing both sides by dt, replacing Π with $C + \alpha S = C - (\partial C / \partial S)S$, and re-arranging, gives

$$\frac{\partial C}{\partial t} = rC - rS\frac{\partial C}{\partial S} - \frac{1}{2}\sigma^2 S^2 \frac{\partial^2 C}{\partial S^2}. \tag{5.30}$$

This is a partial differential equation for the option price C, and is called the *Black–Scholes* equation. To obtain the option price C, one solves this equation.

5.2.3 Creating a portfolio that is equivalent to an option

To derive the Black–Scholes equation, we used the option and the underlying asset to create a portfolio that had no stochastic part, and was therefore equivalent to money. However, we can alternatively construct a portfolio from an amount of money and some of the underlying asset that is equivalent to the option. This allows us to artificially create the option. To do this we need to construct a portfolio whose change in time interval dt is the same as that of the option. To work out what this portfolio needs to be, all we have to do is rearrange the portfolio with value Π that we have already constructed. We know that

$$dC = d\Pi - \alpha dS \tag{5.31}$$

and we know that when $\alpha = -\partial C / \partial S$, then $d\Pi$ is just like having money. So let's make a portfolio out of M dollars and $-\alpha$ of the asset S. Denoting the value of this portfolio by Φ, we have

$$\Phi = M - \alpha S \tag{5.32}$$

and

$$d\Phi = dM - \alpha dS. \tag{5.33}$$

Now, the change in this portfolio, $d\Phi$, will be equal to dC when we choose the amount of money to be $M = \Pi = C - \alpha S$. So to construct a portfolio that will change its value in the next time increment dt by exactly the same amount as the option, we first calculate the current value of the option C by solving the Black–Scholes equation. We then purchase αS worth of the asset and put $M = C - \alpha S$ dollars in the bank (so as to earn the risk-free interest rate). Note that to do this we have to use a total of C dollars, which is just the current value of the option. This is expected, because in order to make a portfolio that is equivalent to the option C, it should have the value of C at every time, and we should therefore have to spend this amount of money to obtain it. If we now wait a time dt, we know that the new value of the portfolio, $\Phi(t + dt)$ still has the same value as the option, $C(t + dt)$, because the change in the two are the same.

To make our portfolio continue to match the value of the option after the next time-step dt, so that $\Phi(t + 2dt) = C(t + 2dt)$, we must change the portfolio somewhat. The reason is that to make this work we had to chose $\alpha = -\partial C/\partial S$, and when we move one time-step to $t + dt$, not only will the value of C change, but the value of the derivative $\partial C/\partial S$ will change as well. Now that we have moved one time-step, we will therefore have to change the value of α, which means that we will have to change the amount of stock we have in our portfolio. Since we know that our portfolio has the value of the option at the present time, $C(t + dt)$, and since we know that to create the correct portfolio we need precisely $C(t + dt)$ dollars, we do not need to add or subtract any money from the portfolio, all we have to do is to "rebalance" the portfolio so that the portion of the total value that is invested in the underlying asset is $(\partial C(t + dt)/\partial S)S$. Of course, for this to work, there needs to be no extra costs associated with the act of purchasing and selling shares (so-called "transaction costs"). If there are some transaction costs, then it will cost us more to create a portfolio equivalent to the option than the "true" value of the option, because we would have to spend this extra money as time went by to periodically rebalance the portfolio.

It is this process of artificially creating an option that allows financial institutions to sell options to people. They sell the option to someone for its current value C, and then use the money to construct the portfolio that is equivalent to the option. They then rebalance the portfolio at intervals to keep it equivalent to the option (they can't do this exactly, but they do it well enough that their risk remains within acceptable bounds). Then, at the time the option matures, if the option is worth something (if it is "in the money") the institution will have exactly this amount in the portfolio, and can give it to the owner of the option. Since constructing the artificial option costs the institution no more than the price that the owner paid for the option, the institution is taking on only minimal risk when it sells the option. However, because of transaction costs, and the work that the institution does in rebalancing the portfolio, it will charge a bit more for the option than its "true" value C.

5.2.4 The price of a "European" option

We now discuss the solution of the Black–Scholes equation for European options. The Black–Scholes equation is a *partial* differential equation for the option price C. This means that C is a function of more than one variable (in this case the two variables S and t) and that the differential equation contains partial derivatives with respect to all of these variables. In this case we call C the *dependent* variable, and S and t the *independent* variables. The differential equations that we have dealt with so far in this course have involved derivatives (actually, differentials) of only

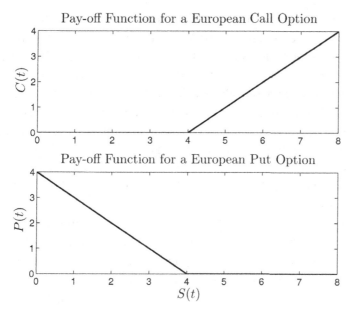

Figure 5.1. The pay-off functions for European call and put options with the strike price $E = 4$.

one independent variable, time. In this case the solution for y at time t depends on the initial value of y, or, more generally, on the value of y specified at any single instant of time.

To obtain a solution to the partial differential equation for C, where the independent variables are t and S, we must specify the value of C at an initial time $t = 0$ (or, in fact, at *any* single time $t = T$), and for *all* values of S at this initial time. In this case, we call these specified values the *boundary* conditions, rather than just the *initial* conditions. Thus to obtain the solution for the value, C, of an option at the present time, we need to know the value of the option at *some* specified time. Fortunately it is easy for us to work out the value of the option at the maturity (or expiry) time T.

The value of the option at the time of maturity, as a function of the share price, is called the *pay-off* function. For a European call option for which the strike price is E, then at the time the option matures, it is worth $S(T) - E$ if $S(T) > E$ (where S is the share price, and T is the maturity time), or zero if $S(T) < E$. We can therefore write the pay-off function for a European call as $f(S) = \max(S(T) - E, 0)$. For a European put (which entitles the bearer to sell the shares for a fixed price E), the pay-off function is $F(S) = \max(E - S(T), 0)$. We plot these pay-off functions in Figure 5.1.

In addition to specifying the values of C at a particular time, we also need to specify the values it takes at the boundaries (the extreme values) of the S variable.

These are $S = 0$ (the shares have no value) and $S \to \infty$ (being the limit in which the share price is much larger than the strike price). By examining the stochastic equation for S, we see that when $S = 0$ is will remain zero for all later times. This means that when $S = 0$ the value of the call option is zero. When S tends to infinity then the value of the call option also tends to infinity. For the put option we have to think a little bit harder. When $S = 0$ at some time t, since we know it will remain zero, we know that the value of the put option at maturity will be the strike price E. If we know that we will have E dollars at time T, then this is the same as having $e^{-(T-t)}E$ dollars in the bank at time t. Thus, on the $S = 0$ boundary, the put option has the value $P(0, t) = e^{-(T-t)}E$. When $S \to \infty$, then $S > E$ so that the value of the put is 0.

Armed with the values of the European call and put options at the final time T, and on the boundaries $S = 0$ and $S \to \infty$, one can solve the Black–Scholes equation to obtain the value of these options at any earlier time t, including the current time $t = 0$ (being the time at which the option is sold). We will not discuss the method by which the Black–Scholes equation is solved, but merely present the solutions. The solution for the European call is

$$C(S, t) = SD(x) - Ee^{-r(T-t)}D(y), \tag{5.34}$$

where D is the distribution function for a Gaussian random variable:

$$D(x) = \frac{1}{\sqrt{2\pi}} \int_{-\infty}^{x} e^{-z^2/2} \, dz, \tag{5.35}$$

and x and y are given by

$$x = \frac{\log(S/E) + (r + \sigma^2/2)(T - t)}{\sigma\sqrt{T - t}} \tag{5.36}$$

and

$$y = \frac{\log(S/E) + (r - \sigma^2/2)(T - t)}{\sigma\sqrt{T - t}}. \tag{5.37}$$

The value of the European put is

$$P(S, t) = -SD(-x) + Ee^{-r(T-t)}D(-y). \tag{5.38}$$

It is important to note that the Black–Scholes equation is not specific to options. To derive the Black–Scholes equation we used only the fact that the value of C was a function of the price of the share S and time t. Thus *any* portfolio whose value V is a function only of S and t will satisfy the Black–Scholes equation. The value of the portfolio at all times (being the solution of the Black–Scholes equation) is therefore determined completely by the value of the portfolio at a specified final time, and at the other "boundaries" $S = 0$ and $S \to \infty$. Thus *any* portfolio that consists only of money, some amount S of shares, and some amount of options on these shares will satisfy the Black–Scholes equation. As a result, two portfolios of this kind whose values are equal to each other for all S at a *single* time T, will have the same value for *all* time, and therefore be *equivalent* portfolios.

5.3 Modeling multiplicative noise in real systems: Stratonovich integrals

In the example of Brownian motion discussed above, the noise did not depend on the state of the particle (the position or velocity of the particle). Noise that is independent of the state of a system is said to be "additive". How do we describe a fluctuating force that is proportional to the position of the particle? Having already treated multiplicative noise such as this in Chapter 3, we are tempted to say that the noise would be described by adding a term proportional to $x\,dW$ to the equation for dp. However, this is not quite correct – there is a subtlety here that must be addressed to get the description right.

The reason for the subtlety is that Wiener noise, the basis of our mathematical treatment of noise, is an idealization. It assumes that in each infinitesimal time interval (no matter how small) the increment due to the noise is *different* (and independent of) the increment in the previous interval. This means that the noise has fluctuations that are infinitely rapid! No physical process can really have this property. To see this another way, consider the average acceleration of the Brownian particle in a given small time interval Δt. This is

$$\bar{a}(\Delta t) = \frac{\Delta p}{\Delta t} = -\gamma p + \beta \frac{\Delta W}{\Delta t}. \qquad (5.39)$$

Now, since the standard deviation of ΔW is $\sqrt{\Delta t}$, the last term is of order $1/\sqrt{\Delta t}$, and thus goes to infinity as Δt goes to zero. Thus the force on the particle is infinite (but fluctuating infinitely fast so that the average over a finite time is finite).

While all real forces will be finite, and thus cannot truly be Wiener noise, as discussed in Section 4.6 the Wiener process is a good approximation to a real fluctuating force so long as this force has fluctuations in a frequency range that is broad compared to the frequency range of the motion of the system. This is why the stochastic differential equations we have been studying are useful for describing physical systems. However, it turns out that there is more than one way to define a stochastic difference equation driven by random increments ΔW, and thus more than one way to take the continuum limit to obtain a stochastic differential equation.

The general Ito difference equation is

$$\Delta x = f(x, t)\Delta t + g(x, t)\Delta W. \qquad (5.40)$$

Now, if we define $x_n \equiv x(n\Delta t)$, $t_n \equiv n\Delta t$ and the first random increment ΔW as ΔW_0, then we can write the equation as

$$x_n = x_{n-1} + \Delta x_{n-1}$$

$$= x_{n-1} + f(x_{n-1}, t_{n-1})\Delta t + g(x_{n-1}, t_{n-1})\Delta W_{n-1}. \tag{5.41}$$

The general solution (obtained by repeatedly adding Δx_n to the initial value of x, $x_0 \equiv x(0)$) is given by

$$x_N = x(N\Delta t) = x_0 + \sum_{n=0}^{N-1} f(x_n, t_n)\Delta t + \sum_{n=0}^{N-1} g(x_n, t_n)\Delta W_n. \tag{5.42}$$

This expression is not very helpful for finding an explicit solution for $x(N\Delta t)$, because to do we so we would have to repeatedly substitute in the solutions for x_{N-1}, x_{N-2}, etc. This would give an unwieldy expansion, ultimately with an infinite number of terms if we tried to take the continuum limit $N \to \infty$. However, this does show us how the Ito integral, defined as

$$\int_0^t g(x(t), t)\, dW \equiv \lim_{N \to \infty} \sum_{n=0}^{N-1} g(x_n, t_n)\Delta W_n, \tag{5.43}$$

appears in the solution for $x(t)$. The subtle point alluded to above is that there is more than one way to define a stochastic integral of the Wiener process. Another way, referred to as the *Stratonovich* integral, is

$$\oint_0^t g(x(t), t)\, dW \equiv \lim_{N \to \infty} \sum_{n=0}^{N-1} g\left(\frac{x[(n+1)\Delta t] + x[n\Delta t]}{2}, n\Delta t\right) \Delta W_n$$

$$\equiv \lim_{N \to \infty} \sum_{n=1}^{N} g\left(\frac{x_{n+1} + x_n}{2}, t_n\right) \Delta W_n, \tag{5.44}$$

where we have written the integral with an "*S*" so as not to confuse it with the Ito integral. Now, if dW was the increment of a nice smooth function of time the Ito and Stratonovich integrals would be equal, but this is not the case. Hence we appear to have an ambiguity when defining the integral of the Wiener process – which one should we use?

Before we answer that question, we will calculate the precise relationship between the two kinds of stochastic integrals, the Ito integral – which is what we have been using in solving our stochastic equations – and the Stratonovich integral. To understand where the difference comes from, note that in the Ito integral, in the nth time-step, it is the value of $g(x, t)$ at the *start* of the time-step that multiplies the increment ΔW_n. The value of g at this time is a function of all the *previous* increments, ΔW_{n-1}, ΔW_{n-2}, etc., but *not* of the new increment ΔW_n. Thus $(\Delta W_n)^2$ does not appear in the nth term. However, in the Stratonovich integral, the nth term contains the value of g at the *end* of the interval for that time-step. This means that in each time-step, the Stratonovich integral contains an extra term proportional to $(\Delta W_n)^2$. As we mentioned above, this would not matter if $W(t)$ was a sufficiently smooth function (that is, differentiable), because $(\Delta W_n)^2$ would tend to zero in the continuum limit. However, the square of the Wiener increment does not vanish in this limit, but is instead equal to dt. It is this that results in the difference between the Ito and Stratonovich integrals. We now calculate precisely what this difference is.

We first define x_n as the solution to the Ito equation

$$x_n = x_{n-1} + f(x_{n-1}, t_{n-1})\Delta t + g(x_{n-1}, t_{n-1})\Delta W_{n-1}. \tag{5.45}$$

By referring to Eq. (5.45) as an Ito equation we mean that x_n is the solution to this stochastic equation where the solution is given by using the Ito definition of a stochastic integral, defined in Eq. (5.43). Next we note that the nth term in the Stratonovich integral is

$$g\left(\frac{x_{n+1} + x_n}{2}, t_n\right)\Delta W_n = g\left(x_n + \frac{\Delta x_n}{2}, t_n\right)\Delta W_n. \tag{5.46}$$

We now expand g as a power series about the point (x_n, t_n). To make the notation more compact we define $g_n \equiv g(x_n, t_n)$. Taking this expansion to second order gives

$$g\left(x_n + \frac{\Delta x_n}{2}, t_n\right) = g_n + \left(\frac{\Delta x_n}{2}\right)\frac{\partial g_n}{\partial x} + \left(\frac{\Delta x_n}{2}\right)^2 \frac{1}{2}\frac{\partial^2 g_n}{\partial x^2}. \tag{5.47}$$

As always, the reason we take this to second order is because Δx_n has a ΔW_n lurking in it. Now we use the fact that $\Delta x_n = f(x_n, t_n)\Delta t + g(x_n, t_n)\Delta W_n$, and remembering that $\Delta W^2 = \Delta t$ this gives

$$g\left(x_n + \frac{\Delta x_n}{2}, t_n\right) = g_n + \left(\frac{f_n}{2}\frac{\partial g_n}{\partial x} + \frac{g_n^2}{4}\frac{\partial^2 g_n}{\partial x^2}\right)\Delta t + \left(\frac{g_n}{2}\frac{\partial g_n}{\partial x}\right)\Delta W_n. \tag{5.48}$$

Now we have expressed the g that appears in the nth term in the Stratonovich integral entirely in terms of the g_n that appears in the Ito integral. Substituting this into the discrete sum for the Stratonovich integral will immediately allow us to write this sum in terms of that for the Ito integral. The result is

$$\sum_{n=0}^{N-1} g\left(\frac{x_{n+1} + x_n}{2}, t_n\right)\Delta W_n = \sum_{n=0}^{N-1} g_n \Delta W_n + \sum_{n=0}^{N-1} \frac{g_n}{2}\frac{\partial g_n}{\partial x}\Delta t. \tag{5.49}$$

Note that in doing this we have dropped the terms proportional to $\Delta t \Delta W_n$. Recall that this is fine because, like $(\Delta t)^2$, they vanish in the continuum limit. The first sum on the right hand side is the familiar sum for the Ito integral, so taking the continuum limit we have

$$\oint_0^t g(x(t), t) \, dW = \int_0^t g(x(t), t) \, dW + \frac{1}{2} \int_0^t \frac{\partial g(x(t), t)}{\partial x} g(x(t), t) \, dt. \quad (5.50)$$

We now see that if we define our solution to the stochastic equation

$$dx = f(x, t)dt + g(x, t)dW \quad (5.51)$$

using the Stratonovich integral instead of the Ito integral, then this solution, which we will call $y(t)$, would be

$$y(t) = y(0) + \int_0^t f(y(t), t) \, dt + \oint_0^t g(y(t), t) \, dW$$

$$= y(0) + \int_0^t \left[f(y(t), t) + \frac{g(y(t), t)}{2} \frac{\partial g(y(t), t)}{\partial y} \right] dt + \int_0^t g(y(t), t) \, dW.$$

But this would be the solution we get if we used the Ito integral to solve the stochastic equation

$$dy = \left[f + \frac{g}{2} \frac{\partial g}{\partial y} \right] dt + g dW. \quad (5.52)$$

So we see that using the Stratonovich integral as the solution to a stochastic differential equation is the same as solving the equation using our usual method (the Ito integral) but changing the deterministic term f by adding $(g/2)\partial g/\partial x$.

Naturally we can also derive the relationship between Ito and Stratonovich stochastic *vector* equations. Let us write the general vector Stratonovich equation

$$d\mathbf{x} = \mathbf{A}(\mathbf{x}, t)dt + B(\mathbf{x}, t)d\mathbf{W} \quad (5.53)$$

explicitly in terms of the elements of $\mathbf{A}$ and B. Denoting these elements as A_i and B_{ij} respectively, the stochastic equation is

$$dx_i = A_i(\mathbf{x}, t)dt + \sum_{j=1}^N B_{ij}(\mathbf{x}, t)dW_j \quad \text{(Stratonovich)}, \quad (5.54)$$

where the x_i are the elements of $\mathbf{x}$, and the dW_j are the mutually independent Wiener noises that are the elements of $d\mathbf{W}$. If Eq. (5.54) is a Stratonovich equation, then the equivalent Ito equation is

$$dx_i = \left(A_i + \frac{1}{2} \sum_{j=1}^N \sum_{k=1}^N B_{kj} \frac{\partial B_{ij}}{\partial v_k} \right) dt + \sum_{j=1}^N B_{ij} dW_j \quad \text{(Ito)}. \quad (5.55)$$

So why have we been studying the Stratonovich integral? The reason is that when we consider multiplicative noise in a real physical system, it is the Stratonovich integral rather than the Ito integral that appears naturally: it turns out that if we have a real fluctuating force, and take the limit in which the bandwidth of the fluctuations becomes very broad (compared to the time-scale of the system), it is the Stratonovich integral which is the limit of this process, *not* the Ito integral. To treat multiplicative noise in physical systems, we should therefore use the following procedure.

Procedure for describing multiplicative noise in physical systems

When we write down a stochastic equation describing the dynamics of a system driven by physical noise whose magnitude depends on the system's state, this is a Stratonovich stochastic equation. Because Ito equations are much simpler to solve than Stratonovich equations, we transform the Stratonovich equation to an Ito equation before proceeding. Then we can use all the techniques we have learned in Chapters 3 and 4. If the noise is purely additive, then the Stratonovich and Ito equations are the same, and so no transformation is required.

Further reading

Recent work on Brownian motion can be found in [18] and references therein. An application of Brownian motion to chemical transport in cells is discussed in [19]. There are numerous books on option pricing and financial derivatives. Some use sophisticated mathematical language, and some do not. One that does not is *The Mathematics of Financial Derivatives: A Student Introduction* by Wilmott, Howison, and Dewynne [20]. Further details regarding Stratonovich equations can be found in the *Handbook of Stochastic Methods*, by Crispin Gardiner [23], and *Numerical Solution of Stochastic Differential Equations*, by Kloeden and Platen [21].

Exercises

1. The current time is $t = 0$, and the risk-free interest rate is r. Assume that there is a commodity whose current price is P, and that a forward contract for the date T is written for this commodity with forward price $F = e^{rT}P$. Determine an expression for the value of the forward contract, $\mathcal{F}(t)$, as a function of time, given that the price of the commodity as a function of time is $P(t)$.

2. (i) Calculate $\partial C / \partial S$ for the European call option. (ii) Show that the expression for the European call option (Eq. (5.34)) is a solution of the Black–Scholes equation.

3. (i) Show that a very simple solution of the Black–Scholes equation is

$$V(S, t) = aS + be^{rt}, \tag{5.56}$$

where a and b are constants. (ii) What portfolio does this solution represent?

4. In Section 5.2.2 we explained how one could have a *negative* amount of an asset in one's portfolio. (Having a negative amount of an asset is called "shorting" the asset.) Options are also assets, so we can have a portfolio that is

$$V(S, t) = C - P, \tag{5.57}$$

where P is a put option and C is a call option. (i) If the call and put have the same maturity time T and strike price E, draw the pay-off function for the portfolio V. (ii) Plot the value of the portfolio

$$U(S, t) = S - Ee^{r(t-T)}, \tag{5.58}$$

at time T. (iii) How is the value of U at time T related to the value of V at time T, and what does this tell you about the relationship of the two portfolios at the current time $t = 0$?

5. If the stochastic equations

$$dx = pdt + \beta(x^2 + p)\, dV \tag{5.59}$$

$$dp = xdt - \gamma(x^3 + p^2)\, dW, \tag{5.60}$$

are Ito equations, then calculate the corresponding Strotonovich equations. Note that dW and dV are mutually independent Wiener noise processes.

6. The equations of motion for a damped harmonic oscillator are

$$dx = (p/m)dt, \tag{5.61}$$

$$dp = -kxdt - \gamma pdt. \tag{5.62}$$

Consider (i) a harmonic oscillator whose spring constant, k, is randomly fluctuating, and (ii) a harmonic oscillator whose damping rate, γ, is randomly fluctuating. Determine the Ito equations for the oscillator in both cases by first writing down the Stratonovich stochastic equations, and then transforming these to Ito equations.

6

Numerical methods for Gaussian noise

6.1 Euler's method

In Chapter 3 we described essentially all of the stochastic equations for which analytic solutions are presently known. Since this is a very small fraction of all possible stochastic equations, one must often solve these equations by simulating them using a computer (a process referred to as *numerical simulation*). To solve the stochastic equation

$$dx = f(x, t)dt + g(x, t)dW \tag{6.1}$$

in this way, we first choose an initial value for x. We can then approximate the differential equation by the equation

$$\Delta x = f(x, t)\Delta t + g(x, t)\Delta W, \tag{6.2}$$

where Δt is a small fixed value, and ΔW is a Gaussian random variable with zero mean and variance equal to Δt. For each time-step Δt, we calculate Δx by using a random number generator to pick a value for ΔW. We then add Δx to x, and repeat the process, continuing to increment x by the new value of Δx for each time-step.

The result of the numerical simulation described above is an approximation to a single sample path of x. To obtain an approximation to the probability density for x at some time T, we can repeat the simulation many times, each time performing the simulation up until time T, and each time using a different set of randomly chosen values for the ΔWs. This generates many different sample paths for x between time 0 and T, and many different samples of the value of x at time T. We can then make a histogram of the values of x, and this is an approximation to the probability density for $x(T)$. We can also obtain an approximation to the mean and variance of $x(T)$ simply by calculating the mean and variance of the samples we have generated.

The above method works just as well for a vector stochastic equation for a set of variables $\mathbf{x} = (x_1, \ldots, x_N)$, driven by a vector of independent noise increments

$\mathbf{dW} = (dW_1, \ldots, dW_M)$:

$$\mathbf{dx} = \mathbf{f}(\mathbf{x}, t)dt + G(\mathbf{x}, t)\mathbf{dW}, \qquad (6.3)$$

where G is an $N \times M$ matrix. As above the approximation to $\mathbf{dx}$ is given by replacing dt with Δt, and $\mathbf{dW}$ with $\Delta \mathbf{W} = (\Delta W_1, \ldots, \Delta W_M)$.

This method for simulating stochastic differential equations is the stochastic equivalent of Euler's method [22] for deterministic differential equations. Below we will describe a more sophisticated method (Milstein's method) that is more accurate.

6.1.1 Generating Gaussian random variables

To realize the simulation described above, we need to generate Gaussian random variables with zero mean. The following method generates two independent zero mean Gaussian variables with variance $\sigma^2 = 1$ [22]. We first take two random variables, x and y, that are uniformly distributed on the interval $[0, 1]$ (All modern programming languages include inbuilt functions to generate such variables.) We then calculate

$$x' = 2x - 1, \qquad (6.4)$$
$$y' = 2y - 1. \qquad (6.5)$$

These new random variables are now uniformly distributed on the interval $[-1, 1]$. We now calculate

$$r = x'^2 + y'^2. \qquad (6.6)$$

If $r = 0$, or $r \geq 1$, then we return to the first step and calculate new random variables x and y. If $r \in (0, 1)$ then we calculate

$$g_1 = x'\sqrt{-2\ln(r)/r}, \qquad (6.7)$$
$$g_2 = y'\sqrt{-2\ln(r)/r}. \qquad (6.8)$$

The variables g_1 and g_2 are Gaussian with zero mean and unit variance, and mutually independent. If instead we want g_1 and g_2 to have variance c, then we simply multiply them by $\sqrt{c}$.

6.2 Checking the accuracy of a solution

The accuracy of the sample paths, means, and variances that we calculate using Euler's method will depend upon the size of the time-step Δt, and the values of x, f and g throughout the evolution. The smaller the time-step with respect to these values, then the more accurate the simulation. The approximate sample path given by the simulation converges to a true sample path in the limit as the time-step tends to zero. A simple way to check the accuracy of a simulation that uses a particular time-step is to perform the simulation again, but this time with the time-step halved. If the results of the two simulations differ by only a little, then the first simulation can be assumed to be accurate to approximately the difference between them. This is because we expect the result of halving the time-step a second time to change the result much less than it did the first time.

The process of halving the time-step, Δt, of a simulation deserves a little more attention. Note that a given sample path is generated by a given realization of the noise, and this is the set of the ΔWs (chosen at random) for the simulation. Let us say that there are N time-steps in our simulation, and denote each of the N noise increments by ΔW_n, where $n = 0, \ldots, N - 1$. If we wish to halve the time-step, and generate an approximation to the *same* sample path, then we need to generate a set of $2N$ Gaussian random numbers, $\Delta \tilde{W}_m$, that agree with the previous set of N random numbers ΔW_n. What this means is that the sum of the first two stochastic increments $\Delta \tilde{W}_0$ and $\Delta \tilde{W}_1$ must equal the first stochastic increment ΔW_0. This is because the total stochastic increment for the second simulation in the time-period Δt, is $\Delta \tilde{W}_0 + \Delta \tilde{W}_1$, and this must agree with the stochastic increment for the first simulation in the same time-step, in order for the two simulations to have the same noise realization. This must also be true for the second pair, $\Delta \tilde{W}_2$ and $\Delta \tilde{W}_3$, etc. So we require

$$\Delta \tilde{W}_{2n} + \Delta \tilde{W}_{2n+1} = \Delta W_n, \qquad n = 0, 1, \ldots, N - 1. \tag{6.9}$$

An example of two approximations to the same sample path, one with half the time-step of the other, is shown in Figure 6.1. Fortunately it is very easy to generate a set of $\Delta \tilde{W}_m$ for which Eq. (6.9) is true. All one has to do is generate N random numbers r_n with mean zero and variance $\Delta t / 2$, and then set

$$\Delta \tilde{W}_{2n} = r_n \tag{6.10}$$

$$\Delta \tilde{W}_{2n+1} = \Delta W_n - r_n. \tag{6.11}$$

The above procedure allows one to perform two simulations of the same sample path for an SDE with different time-steps. If the difference between the final values of x for the two simulations are too large, then one can halve the time-step again and perform another simulation. One stops when the process of halving the time-step changes the final value of x by an amount that is considered to be small enough for the given application.

By repeatedly halving the time-step, one can also determine how rapidly the simulation converges to the true value of $x(T)$. The faster the convergence the

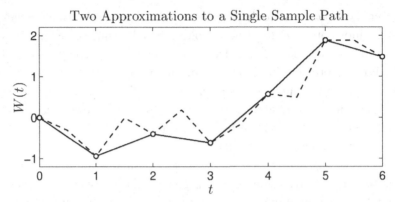

Figure 6.1. The solid line is a numerical approximation to a sample path of the Wiener process, with a time-step of $\Delta t = 1$. The dashed line is an approximation to the same sample path, but with half the time-step. For these approximations to correspond to the same sample path, the second must agree with the first at every multiple of Δt, denoted by the circles.

better, and different numerical methods have different rates of convergence. The simple Euler method that we described above has the slowest rate of convergence.

6.3 The accuracy of a numerical method

The accuracy of a numerical method for simulating an ordinary differential equation, stochastic or otherwise, is measured by how fast the numerical solution converges to the true solution as the time-step, Δt, is reduced. For a given differential equation, and a given sample path, the error in the final value of $x(T)$ scales as some power of the time-step. The higher this power the faster the rate of convergence. The faster the convergence, the larger the time-step that can be used to achieve a given accuracy. The larger the time-step the *fewer* steps are required for the simulation, and (so long as each step is not hugely more complex for the faster method) the simulation can be performed in less time.

In general the error of the Euler method is only guaranteed to scale as Δt (but it sometimes does better – see Section 6.4.1 below). The accuracy of a numerical method is referred to as the *order* of the method, and for stochastic simulations, by convention, this order is quoted as one-half less than the power by which the error scales. The Euler method is thus said to be half-order. There is a perfectly sensible reason for this terminology. It comes from writing the approximate increment as a power series in Δt, which one imagines to agree with the true increment up to some particular power, and then disagree for higher powers. Because the ΔW part of the increment is proportional to $\sqrt{\Delta t}$, one must use a power series in $\sqrt{\Delta t}$, rather than Δt. Thus if the error of the increment Δx is proportional to Δt, as in

Euler's method, then the increment is *accurate* to $\sqrt{\Delta t}$ (half-order in Δt), being the order of the first term in the power series. For a rigorous justification of these arguments, we refer the reader to the comprehensive text on numerical methods for SDEs by Kloeden and Platen [21].

In the next section we describe a simple numerical method that is accurate to first-order in Δt, and thus for which the error scales as $(\Delta t)^{3/2}$.

6.4 Milstein's method

The usefulness of Milstein's method comes from its simplicity, and the fact that it gives a significant improvement over Euler's method, since it is accurate to first-order in Δt. The Milstein method approximates the increment, dx, of the differential equation

$$dx = f(x,t)dt + g(x,t)dW \tag{6.12}$$

by

$$\Delta x = f\Delta t + g\Delta W + \frac{g}{2}\frac{\partial g}{\partial x}\left[(\Delta W)^2 - \Delta t\right]. \tag{6.13}$$

Here we have suppressed the arguments of f and g to avoid cluttering the notation.

6.4.1 Vector equations with scalar noise

For the vector differential equation with a *single* (scalar) noise source, dW, given by

$$d\mathbf{x} = \mathbf{f}(\mathbf{x},t)dt + \mathbf{g}(\mathbf{x},t)dW, \tag{6.14}$$

where $\mathbf{x} = (x_1,\ldots,x_N)^T$, $\mathbf{f} = (f_1,\ldots,f_N)^T$ and $\mathbf{g} = (g_1,\ldots,g_N)^T$, Milstein's method is

$$\Delta x_i = f_i\Delta t + g_i\Delta W + \frac{1}{2}\sum_{j=1}^{M} g_j\frac{\partial g_i}{\partial x_j}\left[(\Delta W)^2 - \Delta t\right]. \tag{6.15}$$

Two special cases of this are as follows.

1. *The Ornstein–Uhlenbeck process with scalar noise.* The equation for this process is $d\mathbf{x} = F\mathbf{x}dt + \mathbf{g}dW$, where F is a constant matrix, and $\mathbf{g}$ is a constant vector. As a result Milstein's method is just the same as Euler's method, being

$$\Delta \mathbf{x} = F\mathbf{x}\Delta t + \mathbf{g}\Delta W. \tag{6.16}$$

This means that for additive noise, Euler's method is a first-order method.

2. *The general linear stochastic equation with scalar noise*. For the general linear stochastic equation with a single noise source, $\mathbf{dx} = F\mathbf{x}\,dt + G\mathbf{x}\,dW$, where F and G are constant matrices, Milstein's method becomes

$$\Delta \mathbf{x} = F\mathbf{x}\,\Delta t + G\mathbf{x}\,\Delta W + \frac{1}{2}G^2\mathbf{x}\left[(\Delta W)^2 - \Delta t\right]. \tag{6.17}$$

6.4.2 Vector equations with commutative noise

When there are *multiple* noise sources (that is, a vector of noise increments) Milstein's method is, in general, considerably more complex. Before we present this method in full, we consider a special case for which the method remains simple. A general vector stochastic differential equation is given by

$$\mathbf{dx} = \mathbf{f}(\mathbf{x}, t)dt + G(\mathbf{x}, t)\mathbf{dW}, \tag{6.18}$$

where $\mathbf{dW}$ is a vector of mutually independent noise sources,

$$\mathbf{dW} = (dW_1, \ldots, dW_M)^{\mathsf{T}}, \tag{6.19}$$

and G is an $N \times M$ matrix whose elements, G_{ij}, may be arbitrary functions of $\mathbf{x}$ and t.

If the matrix G satisfies the set of relations

$$\sum_{m=1}^{N} G_{mj}\frac{\partial G_{ik}}{\partial x_m} = \sum_{m=1}^{N} G_{mk}\frac{\partial G_{ij}}{\partial x_m}, \quad \forall\, i, j, k, \tag{6.20}$$

then the noise is said to be *commutative*. A number of important special cases fall into this category, including *additive noise*, in which G is independent of $\mathbf{x}$, *diagonal noise* in which G is diagonal, and *separated linear noise*, in which

$$G_{ij} = g_{ij}(t)x_i, \tag{6.21}$$

where the $g_{ij}(t)$ are arbitrary functions of time. This last case is referred to as "separated" because G_{ij} does not include x_j, for $j \neq i$.

When the noise is commutative, Milstein's method for solving Eq. (6.18) is

$$\Delta x_i = f_i\,\Delta t + \sum_{j=1}^{M} G_{ij}\,\Delta W_j + \frac{1}{2}\sum_{j=1}^{M}\sum_{k=1}^{M}\left[\sum_{m=1}^{N} G_{mj}\frac{\partial G_{ik}}{\partial x_m}\right]\Delta W_j\,\Delta W_k$$

$$-\frac{1}{2}\sum_{j=1}^{M}\left[\sum_{m=1}^{N} G_{mj}\frac{\partial G_{ij}}{\partial x_m}\right]\Delta t. \tag{6.22}$$

6.4.3 General vector equations

For a completely general vector stochastic equation, the expression for Milstein's approximation to Eq. (6.18) involves a stochastic integral that cannot be written in terms of the discrete noise increments ΔW_j. This expression is

$$\Delta x_i = f_i \Delta t + \sum_{j=1}^{M} G_{ij} \Delta W_j + \sum_{j=1}^{M}\sum_{k=1}^{M}\left[\sum_{m=1}^{N} G_{mj}\frac{\partial G_{ik}}{\partial x_m}\right]\int_t^{t+\Delta t}\int_t^s dW_j(t')dW_k(s).$$

(6.23)

When $j = k$, the double integral does have a simple expression in terms of the discrete stochastic increments, being (see Section 3.8.3)

$$\int_t^{t+\Delta t}\int_t^s dW_j(t')dW_j(s) = \frac{1}{2}\left[(\Delta W_j)^2 - \Delta t\right].$$

(6.24)

But this is not the case when $j \neq k$. Kloeden and Platen suggest a practical method for approximating the double integral for $j \neq k$ [21], and this provides us with a numerical method for solving the vector stochastic equation. To present Kloeden and Platen's approximation we need a few definitions. We define a_{jm}, b_{jm}, and c_{jm}, for each of the integer indices j and m, to be mutually independent Gaussian random variables with zero means and unit variances. We also define

$$\Upsilon_m = \frac{1}{12} - \frac{1}{2\pi^2}\sum_{n=1}^{m}\frac{1}{n^2}.$$

(6.25)

Finally, let us denote the double integral in Eq. (6.23) by

$$H_{jk} \equiv \int_t^{t+\Delta t}\int_t^s dW_j(t')dW_k(s).$$

(6.26)

With these definitions, an approximation to the double integral is given by

$$H_{jk}^{(m)} = \left(\frac{\Delta W_j \Delta W_k}{2} + \sqrt{\frac{\Upsilon_m}{\Delta t}}\left[a_{jm}\Delta W_k - a_{km}\Delta W_j\right]\right)$$
$$+ \frac{\Delta t}{2\pi}\sum_{n=1}^{m}\left(b_{jn}\left[\sqrt{\frac{2}{\Delta t}}\Delta W_k + c_{kn}\right] - b_{kn}\left[\sqrt{\frac{2}{\Delta t}}\Delta W_j + c_{jn}\right]\right).$$ (6.27)

The larger the value of m, the better the approximation to H_{jk}. The average error of the approximation is given by

$$\left\langle\left(H_{jk}^{(m)} - H_{jk}\right)^2\right\rangle = \frac{(\Delta t)^2}{2\pi^2}\sum_{n=m+1}^{\infty}\frac{1}{n^2} \leq \frac{(\Delta t)^2}{2\pi^2 m}.$$

(6.28)

Here the bound is obtained by using the inequality

$$\sum_{n=m+1}^{\infty} \frac{1}{n^2} \leq \int_m^{\infty} \frac{1}{x^2} dx = \frac{1}{m}. \tag{6.29}$$

6.5 Runge–Kutta-like methods

A potential disadvantage of the Milstein method is that one must evaluate the first derivative of the function $g(x, t)$ that multiplies the stochastic increment. For deterministic differential equations it is the Runge–Kutta family of methods that eliminate the need to evaluate such derivatives. Similar methods can be found for stochastic equations. Here we present a first-order method of this type that was obtained by Platen, building upon Milstein's method. We will refer to it as the Milstein–Platen method, and it is obtained from Milstein's method above by replacing the derivative of g with an approximation that is valid to first-order. For a stochastic equation containing only a single variable x, the first-order approximation to the term gg' that appears in Milstein's method is

$$g(x, t)\frac{\partial}{\partial x}g(x, t) \approx \frac{1}{\sqrt{\Delta t}} [g(q, t) - g(x, t)], \tag{6.30}$$

with

$$q = x + f\Delta t + g\sqrt{\Delta t}. \tag{6.31}$$

Substituting this into Milstein's method for a single variable, Eq. (6.13), we obtain the Milstein–Platen method for a single variable:

$$\Delta x = f\Delta t + g\Delta W + \frac{1}{2\sqrt{\Delta t}} [g(q, t) - g(x, t)] [(\Delta W)^2 - \Delta t]. \tag{6.32}$$

For a vector stochastic equation with scalar noise (Eq. (6.14)), the Milstein–Platen method is

$$\Delta x_i = f_i\Delta t + g_i\Delta W + \frac{1}{2\sqrt{\Delta t}} [g_i(\mathbf{q}, t) - g_i(\mathbf{x}, t)] [(\Delta W)^2 - \Delta t], \tag{6.33}$$

where the ith element of the vector $\mathbf{q}$ is

$$q_i = x_i + f_i\Delta t + g_i\sqrt{\Delta t}. \tag{6.34}$$

For a general vector equation, the Milstein–Platen method is obtained by substituting into the Milstein method (Eq. (6.23)) the approximation

$$\sum_{m=1}^{N} G_{mj}(\mathbf{x}, t)\frac{\partial}{\partial x_m}G_{ik}(\mathbf{x}, t) \approx \frac{1}{\sqrt{\Delta t}} [G_{ij}(\mathbf{q}^{(k)}, t) - G_{ij}(\mathbf{x}, t)], \tag{6.35}$$

where the ith element of the vector $\mathbf{q}^{(k)}$ is

$$q_i^{(k)} = x_i + f_i \Delta t + G_{ik}\sqrt{\Delta t}. \tag{6.36}$$

6.6 Implicit methods

Numerical methods for solving differential equations can suffer from instability. An instability is defined as an exponential (and thus rapid) increase in the error, and happens when a small error in one time-step causes a larger error in the next time-step, and so on. This can happen spontaneously at some point in time, even when the solution has been accurate up until that time. While this problem is fairly rare, it is more prevalent for differential equations that generate motion on two or more very different time-scales. (That is, the solution oscillates on a fast time-scale, and also changes on a much longer time-scale, where the longer time-scale is the one we are really interested in.) Differential equations with two or more disparate time-scales are referred to as being "stiff".

If a particular method has a problem with instability, this can usually be fixed by using "implicit" methods. Implicit methods are just like the "explicit" methods we have discussed above, but they use the value of the variable or variables at the *end* of the time-step to evaluate the derivative, instead of the value(s) at the beginning. For example, instead of writing the Euler method for a single variable as

$$x(t + \Delta t) = x(t) + \Delta x(t) = x(t) + f[x(t), t]\Delta t + g[x(t), t]\Delta W, \tag{6.37}$$

we replace $x(t)$ in the functions f and g with $x(t + \Delta t)$. This gives

$$x(t + \Delta t) = x(t) + f[x(t + \Delta t), t]\Delta t + g[x(t + \Delta t), t]\Delta W. \tag{6.38}$$

This method is called "implicit", because it gives an "implicit" equation for $x(t + \Delta t)$ in terms of $x(t)$. To actually calculate $x(t + \Delta t)$ from $x(t)$ we must solve Eq. (6.38) for $x(t + \Delta t)$. This is simple if f and g are linear functions of x, and if not, then we can always use a version of the Newton–Raphson method to solve the implicit equation at each time-step [22].

For a vector stochastic equation (Eq. (6.18)), it turns out that making the replacement $\mathbf{x}(t) \to \mathbf{x}(t + \Delta t)$ in $\Delta \mathbf{x}(t)$ (that is, in the functions $\mathbf{f}$ and G in Eq. (6.18)) in *any* explicit method preserves the order of the method. Thus all the methods we have described above can be immediately turned into implicit methods by making this replacement.

6.7 Weak solutions

So far we have considered the speed at which a numerical method will converge to a sample path. Convergence to a sample path is referred to as *strong* convergence.

Similarly, methods that give strong convergence are called *strong methods*, and the sample paths they generate are called *strong solutions*. If one is not interested in the sample paths, but only in the properties of the probability density of x at some final time T, then one need only consider how fast these properties, such as the mean and variance, converge. These properties are determined by averaging over many sample paths, and for a given numerical method, often converge at a different rate than the individual sample paths. The convergence of the moments of the probability density for a stochastic process $x(T)$ is referred to as *weak* convergence. Methods that provide this kind of convergence are called *weak methods*, and sample paths generated by these methods are called *weak solutions*. Euler's method, described above in Section 6.1, converges at first-order in Δt for the purposes of obtaining weak solutions, so long as the functions $\mathbf{f}(\mathbf{x}, t)$ and $G(\mathbf{x}, t)$ in Eq. (6.3) have continuous fourth derivatives. Thus Euler's method is a half-order strong method, and a first-order weak method.

6.7.1 Second-order weak methods

We can, in fact, obtain second-order weak schemes that are no more complex, and in the general case less complex, than Milstein's strong scheme described above. For a stochastic equation for a single variable (Eq. (6.1)), a second-order method [21] is,

$$\Delta x = f\Delta t + g\Delta W + \frac{1}{2}gg'\left[(\Delta W)^2 - \Delta t\right]$$

$$+ \frac{1}{2}\left(ab' + ba' + \frac{1}{2}b^2b''\right)\Delta W\Delta t$$

$$+ \frac{1}{2}\left(aa' + \frac{1}{2}b^2a''\right)(\Delta t)^2, \tag{6.39}$$

where we have defined $f' \equiv \partial f/\partial x$ and $g' \equiv \partial g/\partial x$.

For a vector stochastic equation (Eq. (6.18)), a second-order method is

$$\Delta x_i = f_i\Delta t + Df_i(\Delta t)^2 + \sum_{j=1}^{M} G_{ij}\Delta W_j$$

$$+ \frac{1}{2}\sum_{j=1}^{M}\left[DG_{ij} + \sum_{k=1}^{N} G_{kj}\frac{\partial f_i}{\partial x_k}\right]\Delta W_j\Delta t$$

$$- \frac{1}{2}\sum_{j=1}^{M}\sum_{k=1}^{M}\left(\sum_{m=1}^{N} G_{mj}\frac{\partial G_{ik}}{\partial x_m}\right)(\Delta W_j\Delta W_k - \xi_{jk}\Delta t), \tag{6.40}$$

where D is the differential operator

$$D = \frac{\partial}{\partial t} + \sum_{n=1}^{N} f_n \frac{\partial}{\partial x_n} + \frac{1}{2} \sum_{n=1}^{N} \sum_{m=1}^{N} \sum_{l=1}^{M} G_{nl} G_{ml} \frac{\partial^2}{\partial x_n \partial x_m}. \tag{6.41}$$

The ξ_{jk} are given by

$$\xi_{ii} = 1, \tag{6.42}$$

$$\xi_{ij} = -\xi_{ji}, \quad \text{for } j < i, \tag{6.43}$$

and for $j > i$ the ξ_{ij} are two-valued independent identical random variables with the probabilities

$$\text{Prob}(\xi_{ij} = \pm 1) = 1/2. \tag{6.44}$$

Further reading

For all the classes of methods we have presented here, there are also higher-order versions, although for strong methods these become rather complex. The interested reader can find all these methods, along with full details of their derivations, in the comprehensive tour de force by Kloeden and Platen [21].

7

Fokker–Planck equations and reaction–diffusion systems

Recall from Chapter 3 that a stochastic equation is a differential equation for a quantity whose rate of change contains a random component. One often refers to a quantity like this as being "driven by noise", and the technical term for it is a *stochastic process*. So far we have found the probability density for a stochastic process by solving the stochastic differential equation for it. There is an alternative method, where instead one derives a partial differential equation for the probability density for the stochastic process. One then solves this equation to obtain the probability density as a function of time. If the process is driven by Gaussian noise, the differential equation for the probability density is called a *Fokker–Planck* equation.

Describing a stochastic process by its Fokker–Planck equation does not give one direct access to as much information as the Ito stochastic differential equation, because it does not provide a practical method to obtain the *sample paths* of the process. However, it can be used to obtain analytic expressions for steady-state probability densities in many cases when these cannot be obtained from the stochastic differential equation. It is also useful for an alternative purpose, that of describing the evolution of many randomly diffusing particles. This is especially useful for modeling chemical reactions, in which the various reagents are simultaneously reacting and diffusing. We examine this application in Section 7.8.

7.1 Deriving the Fokker–Planck equation

Given a stochastic process $x(t)$ with the Ito differential equation

$$dx = f(x, t)dt + g(x, t)dW, \qquad (7.1)$$

then the Fokker–Planck equation can be derived very easily. To do this we first calculate the differential equation for the mean value of an arbitrary function $h(x)$.

Using Ito's rule, the SDE for $h(x)$ is

$$dh = \left(\frac{dh}{dx}\right)f(x,t)dt + \left(\frac{d^2h}{dx^2}\right)\frac{g^2(x,t)}{2}dt + \left(\frac{dh}{dx}\right)g(x,t)dW. \quad (7.2)$$

Taking averages on both sides gives the differential equation for the mean of h, being

$$d\langle h\rangle = \left\langle f(x,t)\left(\frac{dh}{dx}\right)\right\rangle dt + \left\langle \frac{g^2(x,t)}{2}\left(\frac{d^2h}{dx^2}\right)\right\rangle dt, \quad (7.3)$$

or alternatively

$$\frac{d\langle h\rangle}{dt} = \left\langle f(x,t)\left(\frac{dh}{dx}\right)\right\rangle + \left\langle \frac{g^2(x,t)}{2}\left(\frac{d^2h}{dx^2}\right)\right\rangle$$

$$= \int_{-\infty}^{\infty}\left[f(x,t)\left(\frac{dh}{dx}\right) + \frac{g^2(x,t)}{2}\left(\frac{d^2h}{dx^2}\right)\right]P(x,t)dx. \quad (7.4)$$

We now integrate by parts, once for the first term, and twice for the second term. Using the relations $\lim_{x\to\pm\infty}P(x,t) = 0$ and $\lim_{x\to\pm\infty}P(x,t) = 0$, both of which follow from the fact that $\int_{-\infty}^{\infty}P(x,t)dx = 1$, we obtain

$$\frac{d\langle h\rangle}{dt} = \int_{-\infty}^{\infty}h(x)\left\{-\frac{\partial}{\partial x}[f(x,t)P(x,t)] + \frac{1}{2}\frac{\partial^2}{\partial x^2}[g^2(x,t)P(x,t)]\right\}dx. \quad (7.5)$$

We also know that the mean of f is given by

$$\langle h\rangle = \int_{-\infty}^{\infty}h(x)P(x,t)dx, \quad (7.6)$$

so that the derivative of the mean can be written as

$$\frac{d\langle h\rangle}{dt} = \frac{d}{dt}\int_{-\infty}^{\infty}h(x)P(x,t)dx = \int_{-\infty}^{\infty}h(x)\frac{\partial}{\partial t}P(x,t)dx. \quad (7.7)$$

Equating Eqs. (7.5) and (7.7), and realizing that they must be equal for *any* $h(x)$, gives us the Fokker–Planck equation for the probabilty density for $x(t)$:

$$\frac{\partial}{\partial t}P(x,t) = -\frac{\partial}{\partial x}[f(x,t)P(x,t)] + \frac{1}{2}\frac{\partial^2}{\partial x^2}[D(x,t)P(x,t)], \quad (7.8)$$

where we have defined $D(x,t) \equiv g^2(x,t)$.

We can obtain the Fokker–Planck equation for a vector Ito stochastic equation in the same way. For the vector stochastic equation

$$d\mathbf{x} = \mathbf{f}(\mathbf{x},t)dt + G(\mathbf{x},t)d\mathbf{W}, \quad (7.9)$$

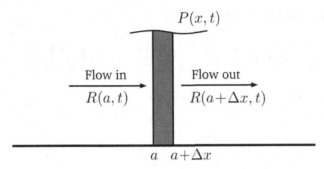

Figure 7.1. A diagram illustrating the calculation of the probability current.

where $\mathbf{x} = (x_1, \ldots, x_N)^{\mathrm{T}}, \mathbf{f} = (f_1, \ldots, f_N)^{\mathrm{T}}, \mathbf{dW} = (dW_1, \ldots, dW_M)^{\mathrm{T}}$ is a vector of M mutually independent Wiener increments, and G is an $N \times M$ matrix, the corresponding Fokker–Planck equation is

$$\frac{\partial}{\partial t} P = -\sum_{i=1}^{N} \frac{\partial}{\partial x_i} [f_i P] + \frac{1}{2} \sum_{i=1}^{N} \sum_{j=1}^{N} \frac{\partial^2}{\partial x_i \partial x_j} [D_{ij} P], \qquad (7.10)$$

where the matrix $D = GG^{\mathrm{T}}$. Here $P \equiv P(\mathbf{x}, t)$ is the joint probability density for the variables $x_1, \ldots, x_N$.

7.2 The probability current

It is useful to write that the Fokker–Planck (FP) equation for a single variable in the form

$$\frac{\partial}{\partial t} P(x, t) = \frac{\partial}{\partial x} \left[-f(x, t) P(x, t) + \frac{1}{2} \frac{\partial}{\partial x} (D(x, t) P(x, t)) \right] \equiv -\frac{\partial}{\partial x} J(x, t),$$

$$(7.11)$$

where $J(x, t)$ is defined as

$$J(x, t) = f(x, t) P(x, t) - \frac{1}{2} \frac{\partial}{\partial x} [D(x, t) P(x, t)]. \qquad (7.12)$$

The relation between P and J, as given by Eq. (7.11) implies that $J(x)$ is the *probability current*: $J(x, t)$ is the rate at which probability is flowing across the point x at time t. To see this, consider the probability that x lies within the narrow interval $[a, a + \Delta x]$. This probability is approximately $P(a, t)\Delta x$. Now consider the rate of change of this probability. As illustrated in Figure 7.1, this rate of change is given by the difference between the rate at which probability is flowing into the interval from the left, and the rate that it is flowing out from the right. Denoting the rate of flow of probability across the point x at time t as $R(x, t)$, we have

$$\frac{\partial}{\partial t}[P(a, t)\Delta x] = R(a, t) - R(a + \Delta x, t).$$
(7.13)

Dividing both sides by Δx, and taking the limit as $\Delta x \to 0$, we get

$$\frac{\partial}{\partial t}P(a, t) = -\lim_{\Delta x \to 0}\frac{R(a + \Delta x, t) - R(a, t)}{\Delta x} = -\frac{\partial}{\partial a}R(a, t).$$
(7.14)

Comparing this with Eq. (7.11) we see that $J(x, t)$ is indeed the rate of flow of the probability across the point x.

When the Fokker–Planck equation contains multiple variables (that is, has more that one spatial dimension), one finds similarly that the probability current vector, $\mathbf{J} = (J_1, \ldots, J_N)$, now giving both the speed and direction of the flow of probability at each point, is given by

$$J_i(\mathbf{x}, t) = f_i P - \frac{1}{2}\sum_{j=1}^{N}\frac{\partial}{\partial x_j}\left(D_{ij}P\right).$$
(7.15)

7.3 Absorbing and reflecting boundaries

To solve an FP equation, one may also need to specify the boundary conditions. If x takes values on the entire real line, then this is unnecessary since we know that P tends to zero as $x \to \pm\infty$, and this will be reflected in the initial condition, being the choice for $P(x, t)$ at $t = 0$. However, if x has some finite domain, say the interval $[a, b]$, then we need to specify what happens at the boundaries a and b. The three most common possibilities are as follows.

1. *Absorbing boundaries.* An absorbing boundary is one in which the particle is removed immediately it hits the boundary. This means that the probability that particle is on the boundary is always zero, and this situation is therefore described by the condition

$$P(c, t) = 0,$$
(7.16)

   where c is the location of the absorbing boundary.
2. *Reflecting boundaries.* A reflecting boundary is one for which the particle cannot pass through. This means that the probability current must be zero across the boundary, and is therefore given by the condition

$$J(c, t) = 0,$$
(7.17)

   where c is the location of the reflecting boundary.
3. *Periodic boundaries.* In this case the two ends (boundaries) of the interval are connected together. This means that the particle is moving on a closed loop such as a circle in one dimension, or a torus in two dimensions. In this case, since the two ends describe the

same physical location, both the probability density and the probability current must be the same at both ends. This is therefore described by the two conditions

$$P(a, t) = P(b, t), \tag{7.18}$$

$$J(a, t) = J(b, t), \tag{7.19}$$

where the interval in which the particle moves is $[a, b]$.

These three kinds of boundary conditions can also be applied to FP equations in more than one dimension. For reflecting boundaries this means setting to zero the dot product of the vector current with the vector normal to the surface of the boundary.

7.4 Stationary solutions for one dimension

When the FP equation is one dimensional (that is, has only a single variable, x), one can fairly easily calculate its *stationary* or *steady-state* solutions. A stationary solution is defined as one in which $P(x, t)$ does not change with time. Often $P(x, t)$ will tend to the stationary solution as $t \to \infty$ for all initial choices of the probability density, and for this reason the stationary solution is important. The differential equation that describes the stationary solutions is obtained by setting $\partial P / \partial t = 0$ in the FP equation. It is therefore

$$-\frac{\partial}{\partial x}[f(x)P(x)] + \frac{1}{2}\frac{\partial^2}{\partial x^2}[D(x)P(x)] = 0 = -\frac{\partial}{\partial x}J(x). \tag{7.20}$$

This equation tells us that the probability current does not change with x, and is therefore the same everywhere. The probability current must also be constant with time, because this is required for P to be constant with time.

If we have reflecting boundary conditions, then $J = 0$ on at least one boundary, and thus $J = 0$ everywhere. In this case the differential equation for the stationary solution is

$$\frac{d}{dx}[D(x)P(x)] = 2f(x)P(x). \tag{7.21}$$

Defining a new function $\xi(x) = D(x)P(x)$, we see that this equation is just a linear differential equation for ξ:

$$\frac{d\xi}{dx} = \left[\frac{2f(x)}{D(x)}\right]\xi. \tag{7.22}$$

The solution is

$$P(x) = \frac{1}{\mathcal{N}D(x)}\exp\left[\int_a^x \frac{2f(u)}{D(u)}du\right], \tag{7.23}$$

where the particle moves in the interval $[a, b]$, and $\mathcal{N}$ is a constant chosen so that

$$\int_a^b P(x)dx = 1. \tag{7.24}$$

If we have periodic boundary conditions, then J will not necessarily vanish. Neither is J a free parameter, however; as we will see, it is completely determined by the assumption of stationarity. In this case the equation for the stationary solution, $P(x)$, is given by

$$\frac{d}{dx}[D(x)P(x)] = 2f(x)P(x) - 2J. \tag{7.25}$$

Once again defining $\xi(x) = D(x)P(x)$, this is a linear equation for ξ, but this time with a constant "driving" term:

$$\frac{d\xi}{dx} = \left[\frac{2f(x)}{D(x)}\right]\xi - 2J. \tag{7.26}$$

The solution (see Section 2.4.1) is

$$P(x) = \left[\frac{Z(x)}{D(x)}\right]\left\{P(a)\left[\frac{D(a)}{Z(a)}\right] - 2J\int_a^x \frac{du}{Z(u)}\right\}, \tag{7.27}$$

where we have defined

$$Z(x) = \exp\left[\int_a^x \frac{2f(u)}{D(u)}du\right]. \tag{7.28}$$

Now we apply the periodic boundary condition $P(a) = P(b)$ (note that we have already applied the boundary condition on J by making J constant), and this gives

$$J = \frac{P(a)}{2\int_a^b \frac{du}{Z(u)}}\left[\frac{D(a)}{Z(a)} - \frac{D(b)}{Z(b)}\right]. \tag{7.29}$$

The solution is therefore

$$P(x) = P(a)\frac{\left[\frac{D(b)}{Z(b)}\int_a^x \frac{du}{Z(u)} - \frac{D(a)}{Z(a)}\int_x^b \frac{du}{Z(u)}\right]}{\frac{D(x)}{Z(x)}\int_a^b \frac{du}{Z(u)}}. \tag{7.30}$$

7.5 Physics: thermalization of a single particle

Consider a physical system whose state is described by the vector $\mathbf{x}$. When this system is placed in contact with a much larger system, called a *bath*, in thermal equilibrium at a temperature T, the probability density for the system "settles down" and becomes stationary. This stationary probability density, $P(\mathbf{x})$, is proportional to $\exp[-\beta E(\mathbf{x})]$, where $E(\mathbf{x})$ is the energy of the system when it is in state $\mathbf{x}$, and $\beta = 1/(k_{\mathrm{B}}T)$, where k_{B} is Boltzmann's constant. The process by which the probability density settles down to a stationary state is referred to as *thermalization*. The stationary probability density is called the *Boltzmann distribution*.

It turns out that the simple stochastic equation that we used to describe Brownian motion in Section 5.1 also provides a model of thermalization for a single particle. We consider a small particle immersed in a fluid, just as in the case of Brownian motion, and as before consider only motion along a single dimension. This means that the state of the system is described by the vector $\mathbf{x} = (x, p)$, so that the FP equation for the system will be two-dimensional. This time we make the description more general than our previous treatment of Brownian motion, by allowing the particle to be subject to an arbitrary spatially dependent force, $F(x)$. In this case the potential energy of the particle is $V(x)$, where $F = -dV/dx$. The stochastic equation for the motion of the particle is now

$$\begin{pmatrix} dx \\ dp \end{pmatrix} = \begin{pmatrix} p/m \\ -V'(x) - \gamma p \end{pmatrix} dt + \begin{pmatrix} 0 \\ g \end{pmatrix} dW, \tag{7.31}$$

where $V'(x) \equiv dV/dx$, and γ is the frictional damping. Note that the total energy of the particle as a function of its state, $\mathbf{x} = (x, p)$, is

$$E(\mathbf{x}) = V(x) + \frac{p^2}{2m}. \tag{7.32}$$

We can now write down the equivalent FP equation for the particle, which is

$$\begin{aligned} \frac{\partial P}{\partial t} &= -\frac{p}{m}\frac{\partial P}{\partial x} + \frac{\partial}{\partial p}\left[V'(x)P + \gamma p P\right] + \frac{g^2}{2}\frac{\partial^2 P}{\partial p^2} \\ &= -\frac{p}{m}\frac{\partial P}{\partial x} + V'(x)\frac{\partial P}{\partial p} + \gamma\frac{\partial}{\partial p}\left[pP + \frac{g^2}{2\gamma}\frac{\partial P}{\partial p}\right]. \end{aligned} \tag{7.33}$$

Substituting

$$P(\mathbf{x}) = \frac{1}{\mathcal{N}}\exp\left[-\left(\frac{\gamma}{g^2}\right)\left(2mV(x) + p^2\right)\right] \tag{7.34}$$

into the FP equation shows that it is the stationary solution. Here, as usual, $\mathcal{N}$ is a constant chosen so that $P(\mathbf{x})$ is normalized. If we choose the strength of the fluctuating force, g, to be

$$g = \sqrt{2\gamma m k_{\mathrm{B}}T} = \sqrt{\frac{2\gamma m}{\beta}}, \tag{7.35}$$

then the stationary solution is exactly the Boltzmann distribution. This shows us that not only does the simple model of Brownian motion in Eq. (7.31) get the statistics of the random motion of the particle correct, but it also gets the thermalizing action of the bath on the particle correct as well! This gives us further reason to trust that Wiener noise is a good approximation to the random force applied to a particle by the molecules of a fluid.

7.6 Time-dependent solutions

Since FP equations describe the same systems as stochastic equations driven by Wiener noise, we already know a number of solutions to FP equations: these are the same as those for the equivalent stochastic equations. In particular, the FP equation for the vector $\mathbf{x} = (x_1, \ldots, x_N)^T$, given by

$$\frac{\partial P}{\partial t} = \sum_{i=1}^{N} \frac{\partial}{\partial x_i} \left[\sum_{j=1}^{N} F_{ij} x_j P \right] + \frac{1}{2} \sum_{i=1}^{N} \sum_{j=1}^{N} D_{ij} \frac{\partial^2 P}{\partial x_i \partial x_j}, \tag{7.36}$$

describes the vector Ornstein–Uhlenbeck process

$$d\mathbf{x} = F\mathbf{x}\,dt + G d\mathbf{W}, \tag{7.37}$$

with $D = GG^T$. Because of this we know that if the initial values of the variables are $\mathbf{x} = \boldsymbol{\mu}_0$, then the solution is the multivariate Gaussian

$$P(\mathbf{x}, t, \boldsymbol{\mu}_0) = \frac{1}{\sqrt{(2\pi)^N \det[\Gamma(t)]}} \exp\left\{-\frac{1}{2}[\mathbf{x} - \boldsymbol{\mu}(t)]^T [\Gamma(t)]^{-1} [\mathbf{x} - \boldsymbol{\mu}(t)]\right\}, \tag{7.38}$$

where the mean vector and covariance matrix are

$$\boldsymbol{\mu}(t) = e^{Ft} \boldsymbol{\mu}_0, \tag{7.39}$$

$$\Gamma(t) = \int_0^t e^{F(t-s)} GG^T e^{F^T(t-s)} ds. \tag{7.40}$$

We can also easily determine the solution if the initial values of the variables are not known precisely. In this case the initial value vector $\boldsymbol{\mu}_0$ is a random variable with a probability density. Since the solution to the stochastic equation for $\mathbf{x}$ is

$$\mathbf{x}(t) = \boldsymbol{\mu}_0 + e^{Ft}\mathbf{x}(0) + \int_0^t e^{F(t-s)} G(s) d\mathbf{W}(s), \tag{7.41}$$

the probability density for $\mathbf{x}(t)$ is the convolution of the probability density for $\boldsymbol{\mu}_0$ with that for $Y(t) \equiv e^{Ft}\mathbf{x}(0) + \int_0^t e^{F(t-s)} G(s) d\mathbf{W}(s)$. The probability density for Y is simply the solution above, Eq. (7.38), with $\boldsymbol{\mu}_0 = \mathbf{0}$. Thus the general solution for a random initial condition is

$$P_{\text{gen}}(\mathbf{x}, t) = \int_{-\infty}^{\infty} \cdots \int_{-\infty}^{\infty} P(\boldsymbol{\mu}_0) P(\mathbf{x} - \boldsymbol{\mu}_0, t, \mathbf{0}) \, d\mu_0^{(1)} \ldots d\mu_0^{(N)}$$

$$= \int_{-\infty}^{\infty} P(\boldsymbol{\mu}_0) P(\mathbf{x}, t, \boldsymbol{\mu}_0) \, d\boldsymbol{\mu}_0. \tag{7.42}$$

Here the elements of $\boldsymbol{\mu}_0$ are denoted by $\mu_0^{(j)}$, $j = 1, \ldots, N$.

7.6.1 Green's functions

There is another way to derive the general solution to the FP equation for the Ornstein–Uhlenbeck process (Eq. (7.42)), and this method can also be applied to other processes. To do this we first note that all FP equations are linear. That is, they do not contain nonlinear functions of P or its derivatives. Because of this, any linear combination of one or more solutions to an FP equation is also a solution. We next note that the solution to an FP equation with a fixed (known) initial value for the variables, $\mathbf{x}(0) = \boldsymbol{\mu}_0$, is a solution for the initial condition

$$P(\mathbf{x}, 0) = \delta(\mathbf{x} - \boldsymbol{\mu}_0). \tag{7.43}$$

Here δ is the Dirac δ-function, described in Section 1.8. Further, using the definition of the δ-function, we can write *any* initial condition for P, $P_0(\mathbf{x})$, as a linear combination (integral) of δ-functions:

$$P_0(\mathbf{x}) = \int_{-\infty}^{\infty} P_0(\mathbf{y}) \delta(\mathbf{x} - \mathbf{y}) \, d\mathbf{y}. \tag{7.44}$$

If the solution to the FP equation for the initial δ-function $\delta(\mathbf{x} - \boldsymbol{\mu}_0)$ is $P_{\boldsymbol{\mu}_0}(\mathbf{x}, t)$, then because all FP equations are linear, the solution for the general initial condition $P(\mathbf{x}, 0) = P_0(\mathbf{x})$ is the *same* linear combination as in Eq. (7.44), but now it is a linear combination of the solutions $P_{\boldsymbol{\mu}_0}(\mathbf{x}, t)$. The general solution is thus

$$P_{\text{gen}}(\mathbf{x}, t) = \int_{-\infty}^{\infty} P_0(\boldsymbol{\mu}_0) P_{\boldsymbol{\mu}_0}(\mathbf{x}, t) \, d\boldsymbol{\mu}_0. \tag{7.45}$$

For the Ornstein–Uhlenbeck FP equation this is exactly the solution given in Eq. (7.42).

The method of solving a partial differential equation by first finding the solution for a δ-function initial condition, and then obtaining the general solution using Eq. (7.45), is called the method of *Green's functions*. The solution for the initial condition $P(\mathbf{x}, 0) = \delta(\mathbf{x} - \boldsymbol{\mu}_0)$ is called the "Green's function".

7.7 Calculating first-passage times

We can use Fokker–Planck equations to determine the probability densities for "first-passage" times of stochastic processes. Recall from Section 4.2 that the first-passage time of a process is the first time at which a sample path reaches some specified value, a. Alternatively, one can consider the first-passage time for a particle to leave a region. This is the time at which a sample path first hits the boundary of the region. For a one-dimensional process this is the time that the process reaches one of the ends of a given interval $[a, b]$. First-passage times are also referred to as "exit times". Here we consider calculating first-passage times for one-dimensional processes in which the drift and diffusion functions, f and D, are time-independent.

7.7.1 The time to exit an interval

We consider now a process that starts at position $x(0) = y$, where y is in the interval $[a, b]$, and show how to obtain a differential equation for the probability that the exit time, T, is greater than t. This is also the probability that the process *remains inside* the interval for all time up to and including t. We will call this probability $P_{in}(t, y)$. To calculate $P_{in}(t, y)$ we impose absorbing boundary conditions on the FP equation at the ends of the interval, $[a, b]$. With these boundary conditions the integral of the solution to the FP equation, $P(x, t|y, 0)$, over the interval $[a, b]$ gives us the probability that the process is still inside the interval at time t, and thus $P_{in}(t, y)$. That is

$$\text{Prob}(T > t) = P_{in}(t, y) = \int_a^b P(x, t|y, 0)dx. \qquad (7.46)$$

We now note that since the FP equation is time homogenous (because f and D are time-independent), shifting the time origin does not change the solution. Thus

$$P_{in}(t, y) = \int_a^b P(x, 0|y, -t)dx. \qquad (7.47)$$

Written this way, the probability that the exit time is greater than t is a function of the time and place at which the process *starts*.

The Fokker–Planck equations we have considered up until now have described the evolution of the probability density for a process, given that it started at a fixed time, and fixed place. But we could also consider deriving a differential equation for a function of the *initial* position and time, given that the final position and time are *fixed*. That is, a differential equation for the function

$$R(y, t) = P(x, 0|y, -t). \qquad (7.48)$$

Once we have $R(y, t)$, then we can obtain $\mathrm{Prob}(T > t)$, using Eq. (7.47) above. The differential equation for $R(y, t)$ is called the *backwards* Fokker–Planck equation. We will not derive the backwards FP here, as this derivation is rather involved. It can be found in sections 3.4 and 3.6 of Gardiner's *Handbook of Stochastic Methods* [23]. For a process described by the (forwards) FP equation

$$\frac{\partial}{\partial t}P(x, t) = -\frac{\partial}{\partial x}[f(x)P(x, t)] + \frac{1}{2}\frac{\partial^2}{\partial x^2}[D(x)P(x, t)], \qquad (7.49)$$

the corresponding backwards FP equation is

$$\frac{\partial}{\partial t}R(y, t) = -f(y)\frac{\partial}{\partial y}R(y, t) + \frac{1}{2}D(y)\frac{\partial^2}{\partial y^2}R(y, t). \qquad (7.50)$$

We note that it is also common to define $R(y, t)$ with a minus sign in front of the t. In that case, the backwards FP equation is given by putting a minus sign in front of the right-hand side of Eq. (7.50).

Since $P_{\mathrm{in}}(t, y)$ is merely the integral of $R(y, t)$ over x (Eq. (7.47)), it is simple to check that it obeys the same equation as $R(y, t)$. To solve the backwards FP equation for $P_{\mathrm{in}}(t, y)$, we need to know the boundary conditions. Since the process is inside the interval at $t = 0$, we have $P_{\mathrm{in}}(0, y) = 1$. If the process starts on the boundaries, then at $t = 0$ it has already reached the exit (since it is immediately absorbed once it is on one of the boundaries), and thus $P_{\mathrm{in}}(t, a) = P_{\mathrm{in}}(t, b) = 0$ for all $t \geq 0$. The differential equation for $P_{\mathrm{in}}(t, y)$ is thus

$$\frac{\partial}{\partial t}P_{\mathrm{in}}(t, y) = -f(y)\frac{\partial}{\partial y}P_{\mathrm{in}}(t, y) + \frac{1}{2}D(y)\frac{\partial^2}{\partial y^2}P_{\mathrm{in}}(t, y), \qquad (7.51)$$

with the boundary conditions

$$P_{\mathrm{in}}(0, y) = 1, \qquad a < y < b, \qquad (7.52)$$

$$P_{\mathrm{in}}(t, a) = P_{\mathrm{in}}(t, a) = 0, \qquad t \geq 0. \qquad (7.53)$$

Since $P_{\mathrm{in}}(t, y)$ is the probability that the time-to-exit, T, is greater than t, the probability *distribution* for T is

$$D_T(t) = \mathrm{Prob}(0 \leq T \leq t) = 1 - P_{\mathrm{in}}(t, y), \qquad (7.54)$$

and thus the probability density for T is

$$P_T(t) = \frac{\partial}{\partial t}D_T(t) = -\frac{\partial}{\partial t}P_{\mathrm{in}}(t, y). \qquad (7.55)$$

The solution to the backward FP equation therefore gives us all the information about the first-exit time from an interval.

The average exit time

Using the above results, we can obtain a closed-form expression for the *average* time it takes a process to exit an interval. The average first-passage time is

$$\langle T \rangle = \int_0^\infty t P_T(t) dt = - \int_0^\infty t \frac{\partial}{\partial t} P_{in}(t, y) dt = \int_0^\infty P_{in}(t, y) dt. \quad (7.56)$$

Here the last step is given by integrating by parts, and using the relation $t P_{in}(t, y) \to 0$ as $t \to \infty$. This relation follows from the assumption that T has a finite mean. We can now obtain a simple differential equation for $\langle T \rangle$, from the differential equation for $P_{in}(t, y)$. Since $\langle T \rangle$ is a function of the initial position of the process, y, we now write it as $\langle T(y) \rangle$. If we integrate both sides of the Fokker–Planck equation for $P_{in}(t, y)$ from 0 to ∞, we get

$$-1 = -f(y) \frac{\partial}{\partial y} \langle T(y) \rangle + \frac{1}{2} D(y) \frac{\partial^2}{\partial y^2} \langle T(y) \rangle, \quad (7.57)$$

where for the right-hand side we have used Eq. (7.56), and for the left-hand side we have used the fact that $P_{in}(\infty, y) = 0$. This equation can be solved by defining $G(y) = \partial \langle T(y) \rangle / \partial y$, and noting that the resulting equation for G can be rearranged into the form of a first-order differential equation with driving (see Section 2.4.1). Applying the boundary conditions, which are $\langle T(a) \rangle = \langle T(b) \rangle = 0$, the solution is

$$\langle T(y) \rangle = \frac{2}{\Omega_a^b} \left[\Omega_a^y \int_y^b \int_a^x \frac{dx dx'}{Z(x) B(x')} - \Omega_y^b \int_a^y \int_a^x \frac{dx dx'}{Z(x) B(x')} \right], \quad (7.58)$$

where

$$Z(y) = \exp \left[-2 \int_a^y \frac{f(x)}{D(x)} dx \right], \quad (7.59)$$

$$\Omega_x^y = \int_x^y \frac{dx'}{Z(x')}. \quad (7.60)$$

7.7.2 *The time to exit through one end of an interval*

The first-passage time for a process to reach a value a from below is equivalent to the time it takes the process to exit an interval $[c, a]$ through the end at a, when $c = -\infty$. We will denote the time that a process exits the interval $[c, a]$ through a by T_a. We can calculate the probability of exit through a particular end of an interval by integrating the probability current at that end. The probability that a process that starts at $x \in [c, a]$ will exit through a at time t or *later* is given by integrating the probability current at a from t to ∞. Denoting this probability by

$P_{in}(t, x)$, we have

$$P_{in}(t, x) = \text{Prob}(T_a \geq t) = \int_t^\infty J(a, t'|x, 0)dt',\tag{7.61}$$

and the probability current at a is given by

$$J(a, t|x, 0) = f(a)P(a, t|x, 0) - \frac{1}{2}\frac{\partial}{\partial a}[D(a)P(a, t|x, 0)].\tag{7.62}$$

We can now derive a differential equation for $P_{in}(t, x)$ by noting that $P(a, t|x, 0)$ obeys the backwards FP equation in x and t (see Section 7.7.1 above). Because of this, substituting the expression for $J(a, t|x, 0)$ in Eq. (7.62) into the backwards FP equation shows that $J(a, t|x, 0)$ also obeys this equation. We now write down the backwards FP equation for $J(a, t|x, 0)$ and integrate both sides with respect to t. The result is the differential equation for $P_{in}(t, x)$:

$$-f(x)\frac{\partial}{\partial x}P_{in} + \frac{D(x)}{2}\frac{\partial^2}{\partial x^2}P_{in} = \frac{\partial}{\partial t}P_{in} = J(a, t|x, 0).\tag{7.63}$$

The boundary conditions for this equation at $x = c$ and $x = a$ are simple. At $x = a$ the particle exits the interval through a at time t, and thus $P_{in}(t, a) = 1$. At $x = c$ the particle exits the interval through c at time t, so it cannot exit through a and we have $P_{in}(t, c) = 0$. However, we do not know the initial condition, $P_{in}(0, x)$, since this is the total probability that the process will exit through a at some time. We will call this total probability $P_a(x)$. One might think that because we do not know $P_a(x)$, we cannot solve the equation for P_{in}. It turns out that we can because $f(x)$ and $D(x)$ are only functions of x. This allows us to derive a differential equation for $P_a(x)$ from the equation for P_{in}.

The total probability of an exit through one end

To calculate $P_a(x)$ we put $t = 0$ in the equation for P_{in}. Since $P_{in}(x, 0) = P_a(x)$, we get the following differential equation for P_a,

$$-f(x)\frac{d}{dx}P_a + \frac{D(x)}{2}\frac{d^2}{dx^2}P_a = J(a, 0|x, 0),\tag{7.64}$$

where we have replaced the partial derivative with respect to x with an ordinary derivative, since x is now the only independent variable. We know that $J(a, 0|x, 0) = 0$ for $x \neq a$, since the process cannot exit through a at $t = 0$ unless $x = a$. While we do not know the value of $J(a, 0|x, 0)$ at $x = a$, as we will see we do not need it. We do need the boundary conditions for P_a, which are $P_a(a) = 1$ (the process *definitely* exits through a if it starts at $x = a$), and $P_a(c) = 0$ (since the process exits through c if it starts at c, so cannot exit through a). We can now write the differential equation for P_a as

$$-f(x)\frac{d}{dx}P_a + \frac{D(x)}{2}\frac{d^2}{dx^2}P_a = 0 \quad \text{for} \quad c \leq x < a.\tag{7.65}$$

We solve this by first defining $Q(x) = dP_a/dx$, and then using separation of variables to obtain

$$Q(x) = k \exp\left[\int_c^x \frac{2f(u)}{D(u)} du\right] \quad \text{for} \quad c \le x < a. \tag{7.66}$$

Note that since we have integrated from c to x to obtain $Q(x)$, we have not required any knowledge of the differential equation at $x = a$. We now integrate $Q(x)$ over x, once again starting at $x = c$, and this gives us $P_a(x)$:

$$P_a(x) = k \int_c^v Q(v)dv \quad \text{for} \quad c \le x < a. \tag{7.67}$$

We see that the boundary condition $P_a(c) = 0$ is already satisfied. The differential equation has not provided us with the value of P_a at $x = a$, but it *has* given us P_a for x arbitrarily close to a. The boundary condition for $x = a$ is now all we need, which is $P_a(a) = 1$. We can satisfy this boundary condition by dividing the expression for $P_a(x)$ above by the integral over $Q(x)$ from c to a. So we finally obtain

$$P_a(x) = \frac{\int_c^x Q(v)dv}{\int_c^a Q(v)dv}, \quad \text{with} \quad Q(v) = \exp\left[\int_c^v \frac{2f(u)}{D(u)} du\right]. \tag{7.68}$$

The probability density for the exit time

We now have the initial condition for P_{in}, and can therefore solve Eq. (7.63) for P_{in}, be it analytically or numerically. From P_{in} we can obtain the probability density for T_a, which we denote by $P_T^{(a)}(t)$. To do this we note that the probability that the process exits through a after time t, *given* that it does exit through a at some time, is $P_{in}/P_a(x)$. The probability distribution for T_a is then $1 - P_{in}/P_a(x)$, so the probability density for T_a is

$$P_T^{(a)}(t) = \frac{-1}{P_a(x)} \frac{\partial}{\partial t} P_{in}(t, x). \tag{7.69}$$

The mean exit time through one end

We can also calculate the equation of motion for the mean exit time through a (given that the process does exit through a). Using Eq. (7.69) the mean exit time is given by

$$\langle T_a(x)\rangle = \int_0^\infty t\, P_T^{(a)}(t)\, dt = \frac{-1}{P_a(x)} \int_0^\infty t \frac{\partial}{\partial t} P_{in}(x, t)\, dt$$

$$= \frac{1}{P_a(x)} \int_0^\infty P_{in}(x, t)\, dt, \tag{7.70}$$

where in the last step we integrated by parts.

To obtain the differential equation for the mean exit time we return to the differential equation for P_{in}, Eq. (7.63). Integrating both sides of this equation from $t = 0$ to ∞, and using Eq. (7.70), gives us the equation of motion for $\langle T_a(x) \rangle$:

$$-f(x)\frac{d}{dx}[P_a(x)\langle T_a(x)\rangle] + \frac{D(x)}{2}\frac{d^2}{dx^2}[P_a(x)\langle T_a(x)\rangle] = -P_a(x). \quad (7.71)$$

Here we have written the derivatives as ordinary rather than partial, because we now only have a single variable, x. Note that we have already determined $P_a(x)$, which is given by Eq. (7.68). All we need now are the boundary conditions. We can state these for the product $P_a(x)\langle T_a(x)\rangle$. Since $\langle T_a(a)\rangle = 0$, and $P_a(c) = 0$, the boundary conditions are

$$P_a(c)\langle T_a(c)\rangle = P_a(a)\langle T_a(a)\rangle = 0. \quad (7.72)$$

The differential equation for $\langle T_a(x) \rangle$ is not difficult to solve. One defines $S(x) \equiv P_a(x)\langle T_a(x)\rangle$, and notes that the resulting equation for S can rearranged into a linear differential equation with driving. This can then be solved using the method in Section 2.4.1, which we leave as an exercise.

7.8 Chemistry: reaction–diffusion equations

Imagine that we have a single particle suspended in liquid and thus undergoing Brownian motion. We saw in Chapter 5 that this motion was well-approximated by the Wiener process. That is, the equation of motion for the x-coordinate of the position of the particle could be written (approximately) as

$$dx = \sqrt{D}dW \quad (7.73)$$

for some positive constant D. The corresponding Fokker–Planck equation for x is therefore

$$\frac{\partial}{\partial t}P(x,t) = \left(\frac{D}{2}\right)\frac{\partial^2}{\partial x^2}P(x,t). \quad (7.74)$$

The stochastic equation is the same for the y- and z-coordinates as well. If we denote the vector of the position of the particle in three dimensions as $\mathbf{x} = (x, y, z)$, then we can write the joint probability density for the position of the particle in these three dimensions as $P(\mathbf{x}, t)$. The Fokker–Planck equation for P is then

$$\frac{\partial P}{\partial t} = \left(\frac{D}{2}\right)\left[\frac{\partial^2 P}{\partial x^2} + \frac{\partial^2 P}{\partial y^2} + \frac{\partial^2 P}{\partial z^2}\right] = \left(\frac{D}{2}\right)\nabla^2 P. \quad (7.75)$$

The sum of the three second spatial derivatives is called the *Laplacian*. To make the notation more compact, one usually writes this as

$$\nabla^2 \equiv \frac{\partial^2}{\partial x^2} + \frac{\partial^2}{\partial y^2} + \frac{\partial^2}{\partial z^2}, \tag{7.76}$$

so that the equation becomes

$$\frac{\partial}{\partial t} P(\mathbf{x}, t) = \left(\frac{D}{2}\right) \nabla^2 P(\mathbf{x}, t). \tag{7.77}$$

Now imagine that we have *many* particles and that the particles are very small. If we divide the volume of liquid up into tiny cubes of size ΔV, then we have a large number of particles in each cube. In this case, to good approximation, the density of the particles in each cube is simply the value of the probability density $P(\mathbf{x}, t)$ for a single particle, multiplied by the total number of particles, N. The equation of motion for the density of particles is therefore simply the equation of motion for the single-particle probability density $P(\mathbf{x}, t)$. So if we denote the density of particles by $\mathcal{D}(\mathbf{x}, t)$, the equation for $\mathcal{D}(\mathbf{x}, t)$ is

$$\frac{\partial}{\partial t} \mathcal{D}(\mathbf{x}, t) = \left(\frac{D}{2}\right) \nabla^2 \mathcal{D}(\mathbf{x}, t). \tag{7.78}$$

This equation describes not only Brownian particles suspended in liquid, but also the density of molecules of a chemical that is dissolved in a liquid (or indeed the molecules of the liquid itself). In the context of density, Eq. (7.78) is referred to as the *diffusion equation*. (This equation also models the flow of heat, and because of this is also referred to as the *heat equation*.) The parameter D is called the *diffusion coefficient*, and determines how fast the particles diffuse through the liquid.

Now consider what happens when there is more than one kind of molecule suspended in the liquid, and these molecules can *react* to form a third molecule. Let's say that $A(\mathbf{x}, t)$ is the density of the first kind of molecule, $B(\mathbf{x}, t)$ that for the second, and they react together to form a third molecule whose density is $C(\mathbf{x}, t)$. We want to derive an equation of motion for these three densities. To do so we assume that the rate at which a molecule of A and B will react to form a molecule of C is proportional to the densities of both molecules. This is because, in order to react, a molecule of A needs to come into "contact" with a molecule of B. The time it takes a molecule of A to find one of B will be proportional to the number of B molecules in the vicinity of A. It will also be proportional to how *many* A molecules find a B molecule in a given time, and this will be proportional to the number of A molecules in the region in question. Thus the reaction rate in a small volume ΔV centered at position $\mathbf{x}$ will be $\gamma A(\mathbf{x}, t) B(\mathbf{x}, t)$ for some positive constant γ.

The densities of A and B will decrease at the rate $\gamma A(\mathbf{x}, t)B(\mathbf{x}, t)$, and that of C will increase at this rate. The differential equations for the three densities, including diffusion, are

$$\frac{\partial}{\partial t}A(\mathbf{x}, t) = -\gamma A(\mathbf{x}, t)B(\mathbf{x}, t) + \left(\frac{D}{2}\right)\nabla^2 A(\mathbf{x}, t) \tag{7.79}$$

$$\frac{\partial}{\partial t}B(\mathbf{x}, t) = -\gamma A(\mathbf{x}, t)B(\mathbf{x}, t) + \left(\frac{D}{2}\right)\nabla^2 B(\mathbf{x}, t) \tag{7.80}$$

$$\frac{\partial}{\partial t}C(\mathbf{x}, t) = \gamma A(\mathbf{x}, t)B(\mathbf{x}, t) + \left(\frac{D}{2}\right)\nabla^2 C(\mathbf{x}, t). \tag{7.81}$$

This is referred to as a set of *reaction–diffusion* equations, because they describe both diffusion of each reagent, and reaction between the reagents.

Now consider the reaction $2A + B \to C$. For this reaction to occur a single molecule of A must find a molecule of B *and* another molecule of A. As before the rate at which it will find a molecule of B is proportional to the number of molecules (and thus the density) of B. The rate at which it will find another molecule of A is almost proportional to the number of A molecules, but this number is reduced by one because we cannot count the molecule of A that is doing the searching. In most chemical reactions we have a very large number of molecules of each reagent, so we can safely ignore the fact that one molecule of A is missing from the available set, and write the total rate of the reaction as $\gamma A^2 B$ for some constant γ.

More generally, if we have a chemical reaction between reagents A and B, such that M molecules of A and N molecules of B react to produce K molecules of C, then we write the reaction as

$$MA + NB \to KC \tag{7.82}$$

and the rate of this reaction is given (to good approximation) by

$$\text{Rate}[MA + NB \to KC] = \gamma A^M(\mathbf{x}, t)B^N(\mathbf{x}, t) \tag{7.83}$$

for some positive constant γ. The set of reaction–diffusion equations describing this reaction is then

$$\frac{\partial A}{\partial t} = -M\gamma A^M B^N + \left(\frac{D}{2}\right)\nabla^2 A \tag{7.84}$$

$$\frac{\partial B}{\partial t} = -N\gamma A^M B^N + \left(\frac{D}{2}\right)\nabla^2 B \tag{7.85}$$

$$\frac{\partial C}{\partial t} = K\gamma A^M B^N + \left(\frac{D}{2}\right)\nabla^2 C. \tag{7.86}$$

7.9 Chemistry: pattern formation in reaction–diffusion systems

We saw in the previous section that, unlike Fokker–Planck equations, reaction–diffusion equations are nonlinear. It turns out that these equations exhibit quite remarkable behavior, generating complex patterns that may be stationary or moving. In this section we introduce the reader to this phenomena via a specific example of a two-dimensional reaction–diffusion equation. We will also discuss how the behavior can be understood in terms of the dynamics of "fronts" by following the analysis of Hagberg and Meron [24, 25, 26].

The reaction–diffusion system we will examine has two chemicals, whose concentrations we will denote by A and B. Since the system has only two spatial dimensions, we define the position vector as $\mathbf{x} \equiv (x, y)$, and the Laplacian as

$$\nabla^2 \equiv \frac{\partial^2}{\partial x^2} + \frac{\partial^2}{\partial y^2}. \tag{7.87}$$

With this definition, the reaction–diffusion system is

$$\frac{\partial}{\partial t} A = A - A^3 - B + \nabla^2 A, \tag{7.88}$$

$$\frac{\partial}{\partial t} B = \varepsilon(A - \alpha B - \beta) + \delta \nabla^2 B. \tag{7.89}$$

Here ε, α, β, and δ are constants, with the constraints that $\varepsilon > 0$ and $\delta > 0$. Note that δ sets the ratio between the diffusion rates for A and B, and ε sets the ratio of the time-scales for A and B (that is, it scales the relationship between the rates at which A and B change with time).

The first thing to note about these equations (and reaction–diffusion equations in general), is that if A and B are homogeneous (that is, do not change with x or y), then the diffusion terms $\nabla^2 A$ and $\nabla^2 B$ are zero. It is useful to examine what stable steady-state solutions there are for A and B in this case. The steady-states are given by setting the time derivatives to zero, and are therefore determined by the equations

$$A - A^3 - B = 0, \tag{7.90}$$
$$A - \alpha B - \beta = 0. \tag{7.91}$$

Writing B as a function of A in both equations gives

$$B = A - A^3, \tag{7.92}$$

$$B = \frac{A - \beta}{\alpha}. \tag{7.93}$$

The solution(s) to these equations are given by the point(s) at which the cubic curve $f(A) = A^3 - A$ intersects with the straight line $g(A) = (A - \beta)/\alpha$. The case we

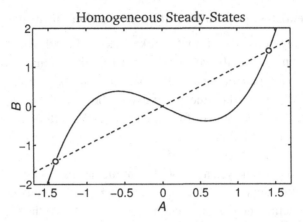

Figure 7.2. The circles mark the two stable homogeneous solutions for the reaction–diffusion system given by Eqs. (7.88) and (7.89), with $\alpha = 1$ and $\beta = 0$.

will consider here is when the parameters α and β are such that these two curves intersect three times, as in Figure 7.2. In this case it turns out that only the two solutions on the outer branches of the cubic are stable. We therefore have two stable solutions that we will label (A_+, B_+) and (A_-, B_-). In the following we will restrict ourselves to the case in which $\beta = 0$. In this case the two solutions are symmetric, so that $(A_-, B_-) = -(A_+, B_+)$.

We consider now a region in which there is a *transition* from one stable solution to the other. In two dimensions this a boundary between two regions (or *domains*) of the two stable homogeneous solutions. Such a boundary is called a *domain wall* or a *front*. For the equivalent one-dimensional system, Hagberg and Meron were able to derive approximate solutions for a domain walls, and to show that these fronts can only move at specific speeds. As we shall see later, this proves very useful in understanding the evolution of fronts in two dimensions. To derive the approximate solutions that describe fronts in one dimension, we first transform to a frame moving at speed v, so that our new x-coordinate is $x' = x - v\tau$. The reaction–diffusion equations become

$$\frac{\partial A}{\partial \tau} = A - A^3 - B + \frac{\partial^2 A}{\partial x'^2} + v\frac{\partial A}{\partial x'} = 0, \tag{7.94}$$

$$\frac{\partial B}{\partial \tau} = \varepsilon(A - \alpha B) + \delta\frac{\partial^2 B}{\partial x'^2} + v\frac{\partial B}{\partial x'} = 0. \tag{7.95}$$

We set the time-derivative to zero, because we assume that the whole front is moving at speed v. This means that the solution will be stationary in the moving frame, and our goal is to determine what values v can take.

To obtain approximate solutions to these equations, for a front at $x' = 0$, we now divide space into three regions: far to the left, where the solution is (A_+, B_+), far to the right where it is (A_-, B_-), and the transition region around $x' = 0$. We also make the approximation that $\mu \equiv \varepsilon/\delta \ll 1$. We skip the details of this procedure, which are given in [25, 26]. When $\beta = 0$ it turns out that the front solutions only exist when the velocity satisfies the equation

$$v = \frac{3v}{q^2\sqrt{2v^2 + 8\varepsilon\delta q^2}}, \qquad (7.96)$$

where we have defined $q^2 = \alpha + 1/2$. For all values of the parameters this equation has the solution $v = 0$, meaning that the front is stationary. Thus in one dimension two domains of different solutions can coexist separated by an unchanging domain wall (front). In two dimensions, if the solution does not change in the y-direction, then the equations reduce to the one-dimensional case. This means that, in two dimensions, two different domains can coexist in the steady-state if they are separated by a stationary domain wall that is a *straight line*. When

$$\delta \geq \delta_{crit} = \frac{9}{8q^6\varepsilon} \qquad (7.97)$$

the stationary solution is the only front solution. However, when $\delta < \delta_{crit}$, there are also the two solutions

$$v = \pm\frac{\sqrt{9 - 8\varepsilon\delta q^6}}{\sqrt{2}q^2}. \qquad (7.98)$$

In this case fronts can move in either direction. In fact, when the two moving solutions are available ($\delta < \delta_{crit}$), the stationary solution becomes unstable. Thus stationary fronts can only be expected to persist for $\delta > \delta_{crit}$.

We consider now a front in two dimensions. We know that if the front is a straight line it will be stationary, and we now ask, what happens to its velocity when it is *curved* slightly? To do this analysis, we assume that the front is a section of a circle with a large radius. (Specifically, we make the radius of curvature much larger than $\sqrt{\delta/\varepsilon}$, which is the length scale upon which B changes across the front.) We choose the center of the circle to be at the origin of the coordinates, and assume that the front is moving in the x-direction, and that its curvature is not changing with time. As before we work in a coordinate system (x', y') moving with the front so that the front is stationary in these coordinates, $x' = x - vt$, and the center of the circle defining the front remains at $(x', y') = (0, 0)$. We focus now on the line $y' = 0$ ($\theta = 0$), which is the x-axis. On this line the front is perpendicular to the x-axis, and thus to the direction of motion. The equations of motion for the front on the line $y' = 0$ are given by Eqs. (7.94) and (7.95).

We now change to radial coordinates, (r, θ), with $x' = r\cos\theta$ and $y' = r\sin\theta$. The location of the front is now specified by $r = r_0$, where r_0 is the radius of

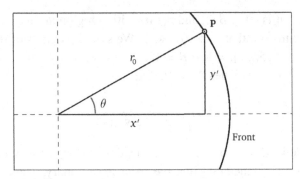

Figure 7.3. Cartesian and radial coordinates for a circular front. The cartesian coordinates of a point **P** on the front are (x', y'), and the radial coordinates are r and θ. The front is defined by $r = r_0$.

curvature of the front (the radius of the circle). The front and the two coordinate systems are shown in Figure 7.3. Note that the reagent concentrations A and B do not change with θ, because they change *across* the front but not *along* the front. This means that the derivatives of A and B with respect to θ are zero. Transforming Eqs. (7.94) and (7.95) to radial coordinates, the equations of motion along the line $y' = 0$, and in the region of the front ($r = r_0$) become

$$A - A^3 - B + \frac{\partial^2 A}{\partial r^2} + (v + \kappa)\frac{\partial A}{\partial r} = 0, \tag{7.99}$$

$$\varepsilon(A - \alpha B) + \delta \frac{\partial^2 B}{\partial r^2} + (v + \delta\kappa)\frac{\partial B}{\partial r} = 0, \tag{7.100}$$

where $\kappa = 1/r_0$ is referred to as the *curvature* of the front. We now multiply the second equation by $(v + \kappa)/(v + \delta\kappa)$, and the equations of motion become

$$A - A^3 - B + \frac{\partial^2 A}{\partial r^2} + (v + \kappa)\frac{\partial A}{\partial r} = 0, \tag{7.101}$$

$$\tilde{\varepsilon}(A - \alpha B) + \tilde{\delta} \frac{\partial^2 B}{\partial r^2} + (v + \kappa)\frac{\partial B}{\partial r} = 0, \tag{7.102}$$

where

$$\tilde{\delta} = \delta \frac{(v + \kappa)}{(v + \delta\kappa)} \quad \text{and} \quad \tilde{\varepsilon} = \varepsilon \frac{(v + \kappa)}{(v + \delta\kappa)}. \tag{7.103}$$

Equations (7.101) and (7.102) are exactly the same as those describing a one-dimensional front, Eqs. (7.94) and (7.95), except that the velocity, v, has been replaced by $v + \kappa$, and the parameters δ and ε have been replaced by $\tilde{\varepsilon}$ and $\tilde{\delta}$. The velocity of the curved front is therefore given by making these replacements in Eq. (7.96). Doing this gives the following implicit equation for the velocity in terms of the curvature:

$$v + \kappa = \frac{3(v + \kappa)}{q^2 \sqrt{2(v + \kappa)^2 \left[1 + \frac{4\varepsilon\delta q^2}{(v+\delta\kappa)^2}\right]}}. \tag{7.104}$$

What we really want to know is whether the velocity *increases* as the curvature increases, or whether it decreases. If it increases then a slight bulge in an otherwise straight front will move faster than the rest of the front, making the bulge *bigger*. This would mean that a straight front is *unstable* to perturbations of the front that are transverse to it; such perturbations will increase over time, and lead to the front becoming increasingly curvy. To determine if this is the case, we make κ sufficiently small that we can expand the implicit expression for the velocity in a power series in κ. If we denote the velocity of a curved front with curvature κ as $v(\kappa)$, and the velocity of a straight front as v_0, then the power series expansion we want is

$$v(\kappa) = v_0 + c_1\kappa + c_2\kappa^2 + \cdots. \tag{7.105}$$

It is quite straightforward to calculate this expansion when $\delta < \delta_{\text{crit}}$ and thus $v_0 > 0$. One finds in this case that c_1 is positive – and thus the front is unstable to transverse perturbations – when $\delta > 3/(\sqrt{8\varepsilon}q^3)$.

It is a little more tricky to calculate the possible values that c_1 can take when we have a stationary front ($\delta > \delta_{\text{crit}}$). We first note that $v = -\kappa$ is always a solution to Eq. (7.104). In this case $c_1 = -1$, and the front is stable. But further investigation reveals that this is not the only solution. To see this we first note that since $v_0 = 0$, $v = c_1\kappa$ to first order in κ. Substituting this expression for v in Eq. (7.104), and simplifying it, we obtain

$$(2\alpha + 1)^3(c_1 + 1)^2\left[(c_1 + \delta)^2\kappa^2 + 2\varepsilon\delta\right] = 18(c_1 + \delta)^2. \tag{7.106}$$

Since we only want the result to first order in κ, we can drop the term proportional to κ^2, and the result is a quadratic equation for c_1. The solutions are

$$c_1 = \frac{-(1 - G\delta) \pm \sqrt{(1 - G\delta)^2 + (1 - G)(G\delta^2 - 1)}}{1 - G}, \tag{7.107}$$

with

$$G = \frac{9}{8q^6\varepsilon\delta}. \tag{7.108}$$

Because we have a stationary front, $\delta > \delta_{\text{crit}}$ and so $(1 - G)$ is positive. This means that there will be a positive solution for c_1 so long as $(G\delta^2 - 1) > 0$, or rather

$$\delta > \left(\frac{8}{9}\right)q^6\varepsilon. \tag{7.109}$$

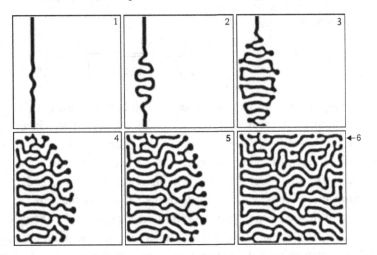

Figure 7.4. The evolution of a reaction–diffusion system that has two homogeneous stationary steady-states. Black denotes one homogeneous steady-state, and white denotes the other. The "labyrinthine" pattern is generated because the domain wall between the two steady-states is unstable to transverse perturbations. This figure is courtesy of Aric Hagberg.

There is therefore a parameter regime in which stationary fronts are unstable to transverse perturbations.

The front instability we have revealed leads to the formation of complex maze-like patterns. Figure 7.4 displays the results of a simulation of this process performed by Hagberg [24, 26]. In frame 1 we start with a front that has some slight wiggles (transverse perturbations). These wiggles extend in frame two, and when the curvature becomes sharp enough, new front lines break off from them (frame 3). These keep extending, forming a maze-like (or "labyrinthine") pattern. This pattern is stable in the steady-state (final frame).

Further reading

Our treatment of Fokker–Planck equations has been mainly limited to one-dimensional problems. A wealth of further information regarding solving Fokker–Planck equations in multiple dimensions, including the method used to obtain the stationary solution to Eq. (7.33), as well as perturbative methods for situations with low noise, are given in the comprehensive text *Handbook of Stochastic Methods* by Crispin Gardiner [23]. We note that the calculation of exit times has applications to transport and binding of chemicals in cells [19]. Further information regarding the behavior of reaction–diffusion systems, and the creation of labyrinthine patterns, can be found in the papers by Hagberg and Meron [24, 25], and Goldstein *et al.* [27].

Exercises

1. Show that the Fokker–Planck equation

$$\frac{\partial P}{\partial t} = -a\frac{\partial P}{\partial x} + \frac{D}{2}\frac{\partial^2 P}{\partial x^2} \tag{7.110}$$

has the solution

$$P(x,t) = \frac{1}{\sqrt{2\pi Dt}}e^{-(x-at)^2/(2Dt)}. \tag{7.111}$$

2. Consider a Fokker–Planck equation that only has diffusion,

$$\frac{\partial P}{\partial t} = \frac{1}{2}\frac{\partial^2}{\partial x^2}[D(x)P], \tag{7.112}$$

and where x is confined to the region $[-1, 1]$ with reflecting boundaries.

(a) Show that the steady-state probability density is inversely proportional to the diffusion $D(x)$. This means that the particle is, on average, pushed away from areas with high diffusion, and confined to areas with low diffusion.

(b) What is the steady-state probability density when $D(x) = 1/\ln(x)$, where k is a positive constant?

(c) What is the steady-state probability density when $D(x) = k(a + |x|)$, where k and a are two positive constants?

(d) What is $P(x)$ in question (c) when $a \to \infty$?

3. Calculate the expression for the probability current in terms of the probability density for each of the two processes

$$dx = a\,dt, \tag{7.113}$$

$$dy = \sqrt{2a(1-y)}\,dW. \tag{7.114}$$

What does this tell you about the effect of a gradient in the diffusion rate?

4. Calculate the Fokker–Planck equation for the stochastic equation

$$dx = a\,dt + bx\,dW, \tag{7.115}$$

where a and b are positive, and x is confined to the interval $[0, 1]$ by reflecting boundaries.

(a) Use the Fokker–Planck equation to determine the steady-state probability density for x. You do not need to determine the normalization constant $\mathcal{N}$.

(b) Calculate the steady-state probability density for $y = 1/x$. This time you need to calculate the normalization constant.

(c) Calculate $\langle 1/x \rangle$ as a function of a and b.

5. Consider a particle whose position, $x(t)$, undergoes the diffusion and damping process

$$dx = -\gamma x\,dt + (1 - x^2)\,dW, \tag{7.116}$$

where x is confined to the interval $[-1, 1]$, and has reflecting boundary conditions at both boundaries.

(a) By inspecting the form of the drift and diffusion functions for x, it is clear that the steady-state probability density will be a symmetric function of x. Explain why.

(b) What is the stationary (steady-state) probability density for x? You do not need to determine the normalization constant $\mathcal{N}$.

(c) Use a computer to plot the steady-state probability density for different values of γ.

(d) Use your plots to guess the form of the steady-state probability density when $\gamma \to 0$.

(e) In view of the answer to (d), what happens to the particle if it is initially placed at $x = 0$ and there is no damping?

6. Show that the probability density given by Eq. (7.34) is a solution to Eq. (7.33).

7. In Section 7.7.2 we derived the differential equation for the total probability that a process exits out of the upper end of an interval $[c, a]$.

(a) Derive the corresponding expression for the total probability, $P_c(x)$, that a process exits out of the lower end, $x = c$, when it starts at a value x in the interval.

(b) Solve the equation for $P_c(x)$.

8. Derive the general solution to Eq. (7.71).

9. Determine the reaction–diffusion equation for the set of simultaneous reactions

$$2A + B \to C, \tag{7.117}$$

$$A + C \to 2B. \tag{7.118}$$

10. Determine the reaction–diffusion equation for the set of simultaneous reactions

$$3A + B \to 2B + C, \tag{7.119}$$

$$A + C \to D. \tag{7.120}$$

11. Derive Eqs. (7.94) and (7.95) from Eqs. (7.88) and (7.89).

12. Derive Eqs. (7.99) and (7.100) from Eqs. (7.94) and (7.95).

8

Jump processes

8.1 The Poisson process

Until now, all our stochastic equations have been driven by the Wiener process, whose infinitesimal increment is denoted by dW. In this chapter we look at stochastic equations driven by a completely different random increment. Recall that the random Wiener increment dW has a Gaussian probability density with mean zero and variance equal to dt. Thus dW can take any real value, although it usually takes values between $\pm 2\sqrt{dt}$. This time our random increment, which we will denote by dN, has only two values: it can be 0 or 1. The probability that dN takes the value 1 in the time interval dt is λdt, and the probability that it takes the value 0 is $1 - \lambda dt$. The probability that dN is 1 in any infinitesimal time interval is therefore vanishingly small, so most of the time dN is zero. However, every now and then dN is 1, and thus the value of $N(t) = \int_0^t dN$ "jumps" by 1. Figure 8.1 shows a sample path of the process $N(t)$, which is called the *Poisson process*.

Let us now work out the probability density for $N(t)$. Note that $N(t)$ is equal to the number of times that $dN = 1$ in the interval $[0, t]$. This is also known as the "number of jumps" in the interval $[0, t]$. Because of this $N(t)$ only takes integer values, and the probability "density" for $N(t)$ is actually a discrete set of probabilities, one for each non-negative integer. We will denote the probability that $N(t) = n$ by $P(n, t)$. We need to work out the value of $P(n, t)$ for each value of n. As usual we discretize the problem, dividing the time interval $[0, t]$ into M equal intervals, each of which has duration $\Delta t = t/M$. This time we will label the M time intervals with the subscript m, where $m = 1, 2, \ldots, M$.

To start with, the probability that there are no jumps, $P(0, t)$, is the product of the probabilities that there is no jump in each interval Δt_m. This is

$$P(0, t) = \lim_{M \to \infty} \prod_{m=1}^{M} (1 - \lambda \Delta t_m) = \lim_{M \to \infty} \left(1 - \lambda \frac{t}{M}\right)^M = e^{-\lambda t}. \qquad (8.1)$$

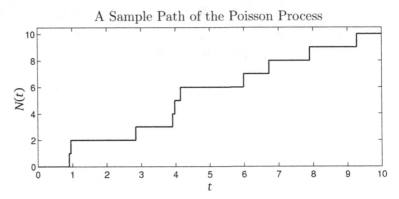

Figure 8.1. A sample path of the Poisson process with jump rate $\lambda = 1$.

The probability that there is exactly one jump in the interval $[0, t]$ is the probability that there is a jump in one of the intervals Δt_m, and no jumps in any of the other intervals. The probability that there is a jump in one *specific* interval, Δt_j, and none in any other interval is the product

$$\lambda \Delta t_j \prod_{m \neq j} (1 - \lambda \Delta t_m) = \lambda \Delta t (1 - \lambda \Delta t)^{M-1}. \tag{8.2}$$

The probability that there is a single jump in the interval $[0, t]$ is the probability that there is a jump in one specific interval, Δt_j, *summed over* all the M intervals in which this jump can occur. It is therefore the above expression multiplied by M. This gives

$$P(1, t) = \lim_{M \to \infty} M \left[\lambda \Delta t (1 - \lambda \Delta t)^{M-1} \right] = \lim_{M \to \infty} \lambda t \left(1 - \lambda \frac{t}{M} \right)^M$$

$$= \lambda t e^{-\lambda t}. \tag{8.3}$$

To calculate the probability that m jumps occur, we similarly calculate the probability that we have m jumps in m specific intervals, and sum this over all the possible distinct sets of m intervals. The probability that there are jumps in a specific set of m intervals is $(\lambda \Delta t)^m (1 - \lambda \Delta t)^{M-m}$. The number of different sets of m intervals in a total of M intervals is an elementary result in combinatorics, and is given by $M!/(m!(M-m)!)$ [28]. So we have

$$P(m, t) = \lim_{M \to \infty} \left[\frac{M!}{m!(M-m)!} \right] (\lambda \Delta t)^m (1 - \lambda \Delta t)^{M-m}$$

$$= \lim_{M \to \infty} \left[\frac{M!}{(M-m)!M^m} \right] \frac{(\lambda t)^m (1 - \lambda t/M)^M}{m!(1 - \lambda t M)^m}. \tag{8.4}$$

We now calculate separately the limits of three parts of this expression:

$$\lim_{M \to \infty} (1 - \lambda t/M)^M = e^{-\lambda t}, \tag{8.5}$$

$$\lim_{M \to \infty} (1 - \lambda t/M)^m = 1, \tag{8.6}$$

$$\lim_{M \to \infty} \frac{M!}{(M-m)! M^m} = \lim_{M \to \infty} \frac{M(M-1)\cdots(M-n+1)}{M^m}$$

$$= \lim_{M \to \infty} \frac{M}{M} \frac{(M-1)}{M} \cdots \frac{(M-n+1)}{M}$$

$$= \lim_{M \to \infty} \frac{M^m}{M^m} = 1. \tag{8.7}$$

With these limits the final result is

$$P(m, t) = \frac{(\lambda t)^m e^{-\lambda t}}{m!}. \tag{8.8}$$

This discrete set of probabilities for the number of jumps, $N(t)$, is called the *Poisson distribution*.

By using the fact that

$$\sum_{m=0}^{\infty} \frac{(\lambda t)^m}{m!} = e^{\lambda t}, \tag{8.9}$$

we see that $P(m, t)$ is correctly normalized:

$$\sum_{m=0}^{\infty} P(m, t) = \sum_{m=0}^{\infty} \frac{(\lambda t)^m e^{-\lambda t}}{m!} = e^{-\lambda t} \sum_{m=0}^{\infty} \frac{(\lambda t)^m}{m!} = e^{-\lambda t} e^{\lambda t} = 1. \tag{8.10}$$

We can also use Eq. (8.9) to calculate the mean and variance of $N(t)$. Setting $\mu = \lambda t$, the mean of $N(t)$ is

$$\langle N(t) \rangle = \sum_{m=0}^{\infty} m P(m, t) = e^{-\mu} \sum_{m=1}^{\infty} \frac{m \mu^m}{m!}$$

$$= e^{-\mu} \sum_{m=1}^{\infty} \frac{\mu^m}{(m-1)!} = \mu e^{-\mu} \sum_{m=1}^{\infty} \frac{\mu^{m-1}}{(m-1)!}$$

$$= \mu e^{-\mu} \sum_{m=0}^{\infty} \frac{\mu^m}{m!} = \mu e^{-\mu} e^{\mu} = \mu = \lambda t. \tag{8.11}$$

We can therefore write the Poisson distribution as

$$P(m) = \frac{\mu^m e^{-\mu}}{m!}, \tag{8.12}$$

where μ is the mean.

The second moment of the Poisson distribution is

$$\langle N^2(t)\rangle = \sum_{m=0}^{\infty} m^2 P(m,t) = e^{-\mu} \sum_{m=0}^{\infty} \frac{m(m-1)\mu^m}{m!} + e^{-\mu} \sum_{m=0}^{\infty} \frac{m\mu^m}{m!}$$

$$= \mu^2 e^{-\mu} \sum_{m=2}^{\infty} \frac{\mu^{m-2}}{(m-2)!} + \langle N(t)\rangle = \mu^2 e^{-\mu} \sum_{m=0}^{\infty} \frac{\mu^m}{m!} + \mu$$

$$= \mu^2 e^{-\mu} e^{\mu} + \mu = \mu^2 + \mu. \tag{8.13}$$

Thus the variance of $N(t)$ is

$$V[N(t)] = \langle N^2(t)\rangle - \langle N(t)\rangle^2 = \mu^2 + \mu - \mu^2 = \mu. \tag{8.14}$$

So the variance of $N(t)$ is equal to the mean.

8.2 Stochastic equations for jump processes

Consider the linear equation

$$dx = \alpha x dt + \beta x dN. \tag{8.15}$$

Since dN takes only the values 0 and 1, it is true that

$$(dN)^2 = dN. \tag{8.16}$$

This is the jump process equivalent of Ito's rule, and much simpler to derive! Further analysis shows that, just like the Wiener process, $dt dN = 0$. We won't derive this result here, but we note that it is to be expected, since

$$\langle dt dN\rangle = dt\langle dN\rangle = \lambda(dt)^2 \equiv 0. \tag{8.17}$$

To solve Eq. (8.15) we proceed in the same way as for the linear Ito stochastic equation: we write the evolution as an exponential. To do this we need to know the expansion for $e^{k dN}$. Using $(dN)^2 = dN$ we have

$$e^{k dN} = \sum_{n=0}^{\infty} \frac{(k dN)^n}{n!} = 1 + dN \sum_{n=1}^{\infty} \frac{k^n}{n!}$$

$$= 1 + dN \left[\sum_{n=0}^{\infty} \frac{k^n}{n!} - 1 \right] = 1 + dN \left[e^k - 1 \right], \tag{8.18}$$

or, alternatively,

$$1 + \beta dN = e^{\ln[1+\beta]dN}. \tag{8.19}$$

Thus

$$x(t + dt) = x + dx = (1 + \alpha dt + \beta dN)x$$
$$= (1 + \alpha dt)(1 + \beta dN)x$$
$$= e^{\alpha dt} e^{\ln[1+\beta]dN} x$$
$$= e^{\alpha dt + \ln[1+\beta]dN} x. \tag{8.20}$$

The solution for $x(t)$ is therefore

$$x(t) = x(0) \exp\left[\alpha \int_0^t ds + \ln[\beta + 1] \int_0^t dN(s)\right]$$
$$= x(0) e^{\alpha t + \ln[1+\beta]N(t)}$$
$$= x(0) e^{\alpha t} (1 + \beta)^{N(t)}. \tag{8.21}$$

Now consider the noise term in Eq. (8.15) in a little more detail. If we have

$$dx = \beta dN \tag{8.22}$$

then the effect of β is merely to change the *size* of the jumps: if $dN = 1$ then $dx = \beta$. Thus in the linear equation above, it is only the size of the jumps that is proportional to x, not the rate of jumps.

If we want to change the rate of the jumps, then we must change λ. Since we cannot directly access the rate of jumps via the stochastic equation, and because in many important problems this rate is a function of time, or even of x, stochastic equations involving dN are of limited use for the purpose of obtaining analytic solutions. To include the rate of jumps explicitly, we need to write an equation of motion for the probability density for the jumping variable. This is the subject of the next section.

8.3 The master equation

Just as we could write a differential equation for the probability density of a stochastic process driven by Wiener noise, we can also write one for a stochastic process driven by the Poisson process. In the case of Wiener noise the equation was called a Fokker–Planck equation, and in this case it is called a *master equation*.

We now derive the master equation for the Poisson process, $N(t)$. Recall that $N(t)$ can take any integer value greater than or equal to zero. As above we will denote the probability that $N(t) = n$ as $P(n, t)$. In each time interval dt, there is a probability λdt that $N(t)$ will jump by 1. Thus, in each time interval dt, the probability that $N(t) = 1$ increases by the probability that $N(t) = 0$, multiplied by λdt, being the probability that there is a jump. Similarly, the probability that

$N(t) = 1$ decreases by the probability that $N(t) = 1$, multiplied by the probability that there is a jump. The equation of motion for each $P(n, t)$ is thus

$$dP(0, t) = -\lambda P(0, n)dt, \tag{8.23}$$

$$dP(n, t) = \lambda P(n - 1, t)dt - \lambda P(n, t)dt, \quad n > 0, \tag{8.24}$$

or

$$\frac{d}{dt}P(0, t) = -\lambda P(0, n), \tag{8.25}$$

$$\frac{d}{dt}P(n, t) = \lambda P(n - 1, t) - \lambda P(n, t), \quad n > 0. \tag{8.26}$$

This is the *master equation* describing the Poisson process.

To solve the master equation we use something called the *generating function*. The generating function is a transform of $P(n, t)$, similar to the characteristic function. It is defined as

$$G(s, t) = \sum_{n=0}^{\infty} s^n P(n, t). \tag{8.27}$$

We now note that

$$\sum_{n=m}^{\infty} s^n P(n - m, t) = s^m \sum_{n=m}^{\infty} s^{n-m} P(n - m, t)$$

$$= s^m \sum_{k=0}^{\infty} s^k P(k, t) = s^m G(s, t), \tag{8.28}$$

and

$$\frac{\partial}{\partial t} G(s, t) = \frac{\partial}{\partial t} \sum_{n=0}^{\infty} s^n P(n, t) = \sum_{n=0}^{\infty} s^n \left[\frac{d}{dt} P(n, t) \right]. \tag{8.29}$$

Using these relations, and summing both sides of the master equation for $P(n, t)$ over n, we obtain the equation of motion for $G(s, t)$:

$$\frac{\partial}{\partial t} G(s, t) = \lambda[sG(s, t) - G(s, t)] = \lambda(s - 1)G(s, t). \tag{8.30}$$

This transforms the master equation for $P(n, t)$ into a partial differential equation (PDE) for $G(s, t)$.

This PDE is not difficult to solve. In fact, since in this case there is no derivative with respect to s, it is actually an ordinary differential equation for G as a function of t. Noting that s is merely a constant as far as t is concerned, we can use separation of variables (see Section 2.4) to obtain the solution, which is

$$G(s, t) = e^{\lambda(s-1)t} G(s, 0). \tag{8.31}$$

Now, $G(s, 0)$ is the initial value of G. This depends on the initial value we choose for $P(n, t)$. If we set the initial value of N to be zero, then $P(n, 0) = \delta_{n0}$, where the symbol δ_{ij} is the very useful *Kronecker delta*, defined by

$$\begin{aligned} \delta_{ij} &= 0, \quad \text{for} \quad i \neq j, \\ \delta_{ij} &= 1, \quad \text{for} \quad i = j. \end{aligned} \tag{8.32}$$

With this choice for $P(n, 0)$, the initial condition for $G(s, t)$ is

$$G(s, 0) = \sum_{n=0}^{\infty} s^n P(n, 0) = \sum_{n=0}^{\infty} s^n \delta_{n0} = 1. \tag{8.33}$$

With this initial condition, the solution is

$$G(s, t) = e^{\lambda(s-1)t}. \tag{8.34}$$

Now we need to calculate $P(n, t)$ from $G(s, t)$. We can do this by expanding G as a Taylor series in s. This gives

$$G(s, t) = e^{\lambda(s-1)t} = e^{-\lambda t} e^{\lambda st} = e^{-\lambda t} \sum_{n=0}^{\infty} \frac{(\lambda t)^n s^n}{n!} = \sum_{n=0}^{\infty} s^n \left(\frac{e^{-\lambda t}(\lambda t)^n}{n!} \right). \tag{8.35}$$

Since $G(s, t) \equiv \sum_{n=0}^{\infty} s^n P(n, t)$, we have

$$P(n, t) = \frac{e^{-\lambda t}(\lambda t)^n}{n!}. \tag{8.36}$$

This is indeed the Poisson distribution that we derived in Section 8.1 using a different method.

8.4 Moments and the generating function

Once one has solved the PDE for the generating function, it is not always possible to obtain a closed-form expression for the set of probabilities $P(n, t)$. Fortunately, we do not need $P(n, t)$ to calculate the mean, variance, and higher moments of $N(t)$; these can be obtained directly from the generating function. To see how this works, consider the derivative of G with respect to s. This is

$$\frac{\partial}{\partial s} G(s, t) = \sum_{n=0}^{\infty} \frac{ds^n}{ds} P(n, t) = \sum_{n=0}^{\infty} n s^{n-1} P(n, t). \tag{8.37}$$

Now, if we put $s = 1$, then we have

$$\frac{\partial}{\partial s} G(s, t) \bigg|_{s=1} = \sum_{n=0}^{\infty} n P(n, t) = \langle N(t) \rangle. \tag{8.38}$$

So $\partial G / \partial s$ evaluated at $s = 1$ gives the mean of N at time t. The second derivative is

$$\frac{\partial^2}{\partial s^2} G(s, t) = \sum_{n=0}^{\infty} \frac{d^2 s^n}{ds^2} P(n, t) = \sum_{n=0}^{\infty} n(n-1) s^{n-2} P(n, t), \tag{8.39}$$

and so

$$\frac{\partial^2}{\partial s^2} G(s, t) \bigg|_{s=1} = \sum_{n=0}^{\infty} n(n-1) P(n, t) = \langle N^2 \rangle - \langle N(t) \rangle. \tag{8.40}$$

The variance of $N(t)$ is therefore

$$V[N(t)] = \frac{\partial^2 G}{\partial s^2} \bigg|_{s=1} + \frac{\partial G}{\partial s} \bigg|_{s=1} - \left(\frac{\partial G}{\partial s} \bigg|_{s=1} \right)^2, \tag{8.41}$$

and higher moments of $N(t)$ can be obtained from higher derivatives of G.

8.5 Another simple jump process: "telegraph noise"

We now consider another simple jump process. In this process our variable $N(t)$ has just two values, $N = 0$ and $N = 1$, and is subject to two random jump processes. The first kind of jump flips the state from $N = 0$ to $N = 1$, and we will call the jump rate for these jumps μ_1. The second kind of jump flips the state back from $N = 1$ to $N = 0$, this time with the jump rate μ_0. This whole process is called "random telegraph noise", a name that comes from the original telegraph that used Morse code for communication. Morse code consists of a series of long and short beeps. Because of this the telegraph line only has two states: when the voltage on the line is low there is no sound at the receiving end, and while the voltage is high the receiving speaker makes a continuous sound. The sender switches between these two states to generate the beeps of Morse code. Noise in the telegraph signal corresponds to random switching between the states due to some disturbance on the line. Hence the name *random telegraph noise*.

The master equation for random telegraph noise is

$$\dot{P}(0, t) = -\mu_1 P(0, t) + \mu_0 P(1, t), \tag{8.42}$$

$$\dot{P}(1, t) = -\mu_0 P(1, t) + \mu_1 P(0, t). \tag{8.43}$$

Since we know that $P(0, t) + P(1, t) = 1$, we can rewrite these equations as separate equations for $P(0, t)$ and $P(1, t)$. This gives

$$\dot{P}(n, t) = -(\mu_1 + \mu_0)P(n, t) + \mu_n, \quad \text{for} \quad n = 0, 1. \tag{8.44}$$

Each of these equations is a linear differential equation with driving, the solution to which is (see Section 2.4.1)

$$P(n, t) = e^{-\gamma t} P(n, 0) + \mu_n \int_0^t e^{\gamma(t'-t)} dt' \tag{8.45}$$

$$= e^{-\gamma t} P(n, 0) + \frac{\mu_n}{\gamma}(1 - e^{-\gamma t}), \tag{8.46}$$

where we have defined $\gamma = \mu_1 + \mu_0$. This is the complete solution for telegraph noise. While this process does not seem very interesting in its own right, with a small modification it will be useful when we look at an application to signals in neurons later in this chapter.

Let us now calculate the auto-correlation function, $\langle N(t)N(t + \tau)\rangle$ for the telegraph process. For this we need the joint probability density that $N = n$ at time t, and that $N = n'$ at time $t + \tau$. To get this, we first note that the joint probability density is given by

$$P(n', t + \tau; n, t) = P(n', t + \tau | n, t)P(n, t), \tag{8.47}$$

where $P(n', t + \tau | n, t)$ is the conditional probability that $N = n'$ at time $t + \tau$, given that $N = n$ at time t.

We now note that if we solve the master equation and choose $N = n_0$ as our initial condition, then the solution is the probability that $N = n$ at time t, *given* that $N = n_0$ at time $t = 0$. That is, the solution is the *conditional* probability density $P(n, t | n_0, 0)$. Putting the relevant initial condition into the general solution, Eq. (8.46), gives

$$P(n, t | n_0, 0) = e^{-\gamma t}\delta_{n,n_0} + \frac{\mu_n}{\gamma}(1 - e^{-\gamma t}), \tag{8.48}$$

where δ_{n,n_0} is the Kronecker delta. We actually want the conditional probability that $N = n'$ at time $t + \tau$, given that $N = n$ at time t. We can get this just as easily by solving the master equation, but setting the initial time equal to t, and evolving for a time τ, so as to get the solution at time $t + \tau$. Using the initial condition $N = n$ at time t, the solution at time $t + \tau$ is thus

$$P(n', t + \tau | n, t) = e^{-\gamma \tau}\delta_{n',n} + \frac{\mu_{n'}}{\gamma}(1 - e^{-\gamma \tau}). \tag{8.49}$$

To obtain $P(n', t + \tau; n, t)$ all we need now is the probability density for N at time t, so that we can use Eq. (8.47). To get $P(n, t)$, we simply solve the telegraph process with the initial condition $N = n_0$ at time $t = 0$. The probability that $N = n$ at time t is then given by Eq. (8.48). So we have finally

$$P(n', t + \tau; n, t) = \left[e^{-\gamma\tau} \delta_{n',n} + \frac{\mu_{n'}}{\gamma} (1 - e^{-\gamma\tau}) \right] \left[e^{-\gamma t} \delta_{n,n_0} + \frac{\mu_n}{\gamma} (1 - e^{-\gamma t}) \right].$$

(8.50)

Now we can easily calculate the auto-correlation function. This is

$$\langle N(t)N(t + \tau) \rangle = \sum_{n'} \sum_{n} n' n P(n', t + \tau; n, t) = P(1, t + \tau; 1, t)$$

$$= \left[e^{-\gamma\tau} + \frac{\mu_1}{\gamma} (1 - e^{-\gamma\tau}) \right] \left[e^{-\gamma t} \delta_{1,n_0} + \frac{\mu_1}{\gamma} (1 - e^{-\gamma t}) \right].$$

(8.51)

Now see what happens as $t \to \infty$. The initial condition, $N = n_0$ becomes irrelevant, and the time t drops out completely. The auto-correlation function is then only a function of the time difference τ, and is

$$g(\tau) = \langle N(t)N(t + \tau) \rangle = \frac{\mu_1}{\gamma} \left[e^{-\gamma\tau} + \frac{\mu_1}{\gamma} (1 - e^{-\gamma\tau}) \right].$$

(8.52)

This is the auto-correlation function in the steady-state, when all initial transients have died away. Now let's check that this is what we expect. When $\tau = 0$, then $N(t)$ and $N(t + \tau)$ are the same, and $g(0)$ is just the expectation value of N^2 in the steady-state. If the jump rates μ_0 and μ_1 are equal, then we would expect there to be a $1/2$ probability of the system being in either state. Thus $g(0) = \langle N^2 \rangle$ should be $1/2$ when $\mu_0 = \mu_1$, and from Eq. (8.52) we see that this is true. When $\tau \to \infty$, we would expect $N(t)$ and $N(t + \tau)$ to be completely independent. Hence we would expect $g(\infty)$ to be equal to the steady-state value of $\langle N \rangle^2$. Putting $\mu_0 = \mu_1$, one has $\langle N \rangle = 1/2$, and indeed $g(\infty) = 1/4$.

8.6 Solving the master equation: a more complex example

Sometimes we might have a process where the rate of jumps depends upon the value of the jumping variable at the current time. Consider a variable N that is subject to a single jump process in which its value increases by 1, as in the Poisson process, but this time the rate of jumps is proportional to the value of N. The master equation for this process is

$$\frac{dP(0, t)}{dt} = 0,$$

(8.53)

$$\frac{dP(n, t)}{dt} = \lambda(n - 1)P(n - 1, t) - \lambda n P(n, t), \quad n > 0.$$

(8.54)

We can work out the equation of motion for the generating function by noting that

$$\frac{\partial}{\partial s}G(s,t) = \sum_{n=0}^{\infty}\left(\frac{d}{ds}s^n\right)P(n,t) = \sum_{n=0}^{\infty}ns^{n-1}P(n,t)$$

$$= \frac{1}{s}\sum_{n=0}^{\infty}ns^n P(n,t). \tag{8.55}$$

The differential equation for $G(s,t)$ is then

$$\frac{\partial}{\partial t}G(s,t) = \lambda(s^2 - s)\frac{\partial}{\partial s}G(s,t). \tag{8.56}$$

We can solve this equation using a "separation of variables" technique for partial differential equations. We first write G as the product of a function of s and a function of t, so that

$$G(s,t) = H(s)L(t). \tag{8.57}$$

Substituting this into the differential equation for G, we can rearrange the equation so that L only appears on the left-hand side, and H only appears on the right-hand side:

$$\frac{1}{\lambda L(t)}\frac{dL(t)}{dt} = \frac{(s^2 - s)}{H(s)}\frac{dH(s)}{ds}. \tag{8.58}$$

The left-hand side depends only on t, and the right-hand side depends only on s, and these two variables are independent. Thus the only way the above equation can be true is if both sides are equal to a constant that is independent of s and t. Denoting this constant by c, the result is two separate differential equations:

$$\frac{dL}{dt} = c\lambda L,$$

$$\frac{dH}{ds} = c\frac{H(s)}{s^2 - s}. \tag{8.59}$$

Both these equations are readily solved using separation of variables, discussed in Section 2.4. The solution to the first is $L = Ae^{c\lambda t}$ for some constant A. The solution to the second is

$$H(s) = B\left(\frac{s-1}{s}\right)^c = B\left(1 - s^{-1}\right)^c \tag{8.60}$$

for some constant B. A solution for $G(s,t)$ is therefore

$$G(s,t) = H(s)L(t) = D\left(1 - s^{-1}\right)^c e^{c\lambda t} = D\left[\left(1 - s^{-1}\right)e^{\lambda t}\right]^c, \tag{8.61}$$

where D and c are arbitrary constants. This is not the most general solution for $G(s,t)$, however. Since the equation for $G(s,t)$ is linear, any linear combination of different solutions of this equation is also a solution. Thus the general solution for G is a sum of as many solutions of the above form as we want, with a different value of D and c for each term in the sum. However, this sum is not the best way to write the general solution. By substituting into the differential equation for $G(s, t)$, it is simple to verify that

$$G(s, t) = F[H(s)L(t)] = F\left[\left(1 - s^{-1}\right) e^{\lambda t}\right] \tag{8.62}$$

is a solution for *any* function F. (This is equivalent to the solution written in terms of the sum described above, a fact which can be seen by using a power series representation for F.)

Now that we have the general solution for the generating function, we can find the specific solution for a given initial condition. Let us choose the simple initial condition $N = m$ at time $t = 0$. This means that $P(m, 0) = 1$, so that $G(s, 0) = s^m$. To satisfy this initial condition we must choose F so that

$$F\left(\frac{s - 1}{s}\right) = G(s, 0) = s^m. \tag{8.63}$$

The process of determining the function F from this equation will probably seem a bit confusing at first. To find F we first note that the functions that map s^m to s, and then s to $(s - 1)/s$ are easy to find. The first is the function $f(x) = x^{1/m}$, and the second is the function $g(x) = (x - 1)/x$. Thus the *inverse* of F is easy to find, as it is just the concatenation of g and f. Defining $x \equiv (s - 1)/s$ and $y \equiv s^m$ we have

$$x = \frac{s - 1}{s} = F^{-1}(s^m) = f[g(s^m)] = f[g(y)]. \tag{8.64}$$

Thus

$$x = f[g(y)] = \frac{y^{1/m} - 1}{y^{1/m}} \tag{8.65}$$

and all we have to do is solve this for y. This gives

$$y = \left(\frac{1}{1 - x}\right)^m, \tag{8.66}$$

and hence

$$F(x) = \left(\frac{1}{1 - x}\right)^m. \tag{8.67}$$

Now we have the solution for the generating function satisfying the initial condition $N(0) = m$. This is

$$G(s, t) = F\left[\left(1 - s^{-1}\right) e^{\lambda t}\right] = \left(\frac{s e^{-\lambda t}}{s(e^{-\lambda t} - 1) + 1}\right)^m. \tag{8.68}$$

This time it is not so easy to obtain an explicit expression for the probabilities for N as a function of time, $P(n, t)$. We therefore use the method of Section 8.4 above to calculate the mean directly from the generating function. We have

$$\frac{\partial G}{\partial s} = m \left(\frac{se^{-\lambda t}}{s(e^{-\lambda t} - 1) + 1} \right)^{m-1} \left(\frac{e^{-\lambda t}}{s(e^{-\lambda t} - 1) + 1} - \frac{se^{-\lambda t}(e^{-\lambda t} - 1)}{[s(e^{-\lambda t} - 1) + 1]^2} \right).$$

(8.69)

Putting $s = 1$ in this expression gives

$$\langle N(t) \rangle = m e^{\lambda t} \tag{8.70}$$

The calculation of the variance, while similar, is rather tedious, and so we wont bother to derive it here.

8.7 The general form of the master equation

In general one may wish to consider a number of discrete variables M_i that are subject to jump processes, and the rates of these jumps may in general be any function of the values of all of these variables. To start simply, let us say that we have a single discrete variable M, and that it is subject to a jump process in which the value of M jumps by n. In general the rate, R, of this jump process can be any function of time t and of the value of M. So we will write the rate as $R(M, t)$.

The master equation for $P(m, t)$, being the probability that $M = m$ at time t, is then

$$\dot{P}(m, t) = R(m - n, t)P(m - n, t) - R(m, t)P(m, t). \tag{8.71}$$

That is, each jump process with jump size n contributes two terms to the right-hand side of the master equation, one corresponding to jumps that take M *to* the value m (from $M = m - n$), and one that takes M *from* the value m. If M can only take non-negative values, then there will also be special cases for $\dot{P}(m)$ when $m < n$.

If we have a single variable, M, that is subject to J different jump processes, each with a jump size of n_j, then the master equation is simply the sum of all the terms from each jump process. Thus the general form of the master equation for a single variable M is

$$\dot{P}(m, t) = \sum_{j=1}^{J} [R_j(m - n_j, t)P(m - n_j, t) - R_j(m, t)P(m, t)], \tag{8.72}$$

where R_j is naturally the rate of jump process j. In general there will also, of course, be a set of special cases for each $m < n_j$.

When we have more than one variable, then the master equation describes the dynamics of a set of *joint* probabilities for the values of the variables. Considering just two variables, M_1 and M_2, is enough to see the general form. We write the joint probability that $M_1 = m_1$ and $M_2 = m_2$, at time t, as $P(m_1, m_2, t)$. If M_1 undergoes a jump process that adds n at each jump, then in general this can depend upon the values of *both* M_1 and M_2. Further, a single jump process can simultaneously change the values of *both* variables. If we have one jump process with rate $R(M_1, M_2, t)$, and this adds n_1 to M_1 and n_2 to M_2, then the master equation is

$$\dot{P}(m_1, m_2, t) = R(m_1 - n_1, m_2 - n_2, t)P(m_1 - n_1, m_2 - n_2, t)$$
$$- R(m_1, m_2, t)P(m_1, m_2, t). \tag{8.73}$$

Once again a single jump process contributes only two terms to the master equation. In the general case, in which we have J jump processes, then once again the master equation is given by summing the terms from all the jump processes. Let us denote the rate of the jth jump process as $R_j(M_1, M_2)$, and the change it induces in the value of M_i by $n_{i,j}$. In writing the master equation it is also convenient to use a more compact notation by defining the vectors $\mathbf{M} = (M_1, M_2)$, $\mathbf{m} = (m_1, m_2)$, and $\mathbf{n}_j = (n_{1,j}, n_{2,j})$. With this vector notation the master equation is

$$\dot{P}(\mathbf{m}, t) = \sum_{j=1}^{J} [R_j(\mathbf{m} - \mathbf{n}_j, t)P(\mathbf{m} - \mathbf{n}_j, t) - R_j(\mathbf{m}, t)P(\mathbf{m}, t)]. \tag{8.74}$$

The stochastic equations corresponding to the above master equation are as follows. For each kind of jump event we need to include one Poisson process. If we denote the Poisson process for the jth jump process as dN_j, then the probability that $dN_j = 1$ in a given time-step dt is $R_j(\mathbf{M})dt$. The stochastic differential equations for M_1 and M_2 are then

$$dM_1 = \sum_{j=1}^{J} n_{1,j} \, dN_j, \tag{8.75}$$

$$dM_2 = \sum_{j=1}^{J} n_{2,j} \, dN_j. \tag{8.76}$$

8.8 Biology: predator–prey systems

A natural application of jump processes is to populations of organisms, such as people, animals, or bacteria. This is because the size of a population is an integer, and the occurrence of a birth or death is a jump of ± 1. Further, because there are many individuals in a population, all making different decisions about when to have children, it seems reasonable to model the occurrence of births and deaths as random events.

A specific example of population dynamics involves two populations, one of which, the *predator*, eats the other, being the *prey*. If we denote the population of the prey by M (mice), and the population of the predator by J (jaguars), then we might model the system with three jump processes.

(1) $M \to M + 1$, at the rate λM, for some constant λ.
(2) $J \to J + 1$ and $M \to M - 1$, at the rate $\mu J M$ for some constant μ.
(3) $J \to J - 1$, at the rate νJ for some constant ν.

Jump process (1) corresponds to the birth of one member of the prey. If we assume that each member of the prey has a child at some average rate, λ, the rate of these jump events should be λM. In choosing λ to be a constant, we are assuming that the population of the prey is not limited by resources; their food supply is inexhaustible. Jump event (2) corresponds to a member of the predators eating one of the prey, and immediately reproducing. This is somewhat simplistic, but it does capture the effect that the amount that the predators have to eat affects how fast their population increases. We choose the rate at which a predator eats one of the prey as being proportional to both the number of predators and the number of prey. This means that the predators find it easier to find and catch the prey if they are more plentiful. Jump event (3) corresponds to one of the predators dying, and the rate at which this happens is naturally proportional to the number of predators. We could, of course, include a fourth jump event, being the death of one of the prey. However, the "classic" predator–prey model does not do this, instead allowing the predators to be the only check on the population of the prey. This is not unreasonable, given that populations of animals invariably increase without bound given a plentiful food supply and an absence of predators.

Given the three jump events, we can now write a master equation for the populations M and J. If we denote the joint probability that $M = m$ and $J = j$ as $P(j, m)$, then the master equation is

$$\dot{P}(j, m) = \lambda \left[(m - 1)P(j, m - 1) - m P(j, m)\right]$$
$$+ \mu \left[(j - 1)(m + 1)P(j - 1, j + 1) - jm P(j, m)\right]$$
$$+ \nu \left[(j + 1)P(j + 1, m) - j P(j, m)\right], \qquad (8.77)$$

with, as usual, special cases for $m = 0$ and $n = 0$:

$$\dot{P}(j, 0) = \mu(j - 1)P(j - 1, 1)$$
$$+ \nu \left[(j + 1)P(j + 1, 0) - j P(j, 0)\right], \qquad (8.78)$$
$$\dot{P}(0, m) = \lambda \left[(m - 1)P(0, m - 1) - m P(0, m)\right] + \nu P(1, m), \qquad (8.79)$$
$$\dot{P}(0, 0) = \nu P(1, 0). \qquad (8.80)$$

Unfortunately there is no known analytic solution to this master equation. We must therefore solve it numerically (that is, by simulating the evolution of the two populations on a computer).

Rather than solving the master equation to obtain the joint probability density $P(j, m, t)$, it is more interesting to solve for a particular realization of the jump process. That is, to choose the random times at which each of the jump events occur by picking random numbers from the correct probability densities. This means that we directly simulate the stochastic equations corresponding to the master equation. The stochastic equations for the predator–prey system are

$$dM = dN_1(M) - dN_2(J, M),$$ (8.81)

$$dJ = dN_2(J, M) - dN_3(J),$$ (8.82)

where the three jump processes, dN_j, have the respective jump rates

$$r_1 = \lambda M,$$ (8.83)

$$r_2 = \mu J M,$$ (8.84)

$$r_3 = \nu J.$$ (8.85)

This means that the respective probabilities that each of the dN_i are equal to 1 in a time-step dt at time t are

$$P(dN_1 = 1) = \lambda M(t)dt,$$ (8.86)

$$P(dN_2 = 1) = \mu J(t)M(t)dt,$$ (8.87)

$$P(dN_3 = 1) = \nu J(t)dt.$$ (8.88)

Performing a numerical simulation of the coupled stochastic equations (8.81) and (8.82) is very simple. One first chooses initial values for the two populations M and J. At each time-step dt, one chooses three independent random numbers, a_i, $i = 1, 2, 3$, each distributed evenly between 0 and 1. For each i, we set $dN_i = 1$ if $a_i \leq P(dN_i = 1)$, and set $dN_i = 0$ otherwise. We then calculate dM and dJ for that time-step using Eqs. (8.81) and (8.82). We repeat this for many small time-steps, and the result is a single realization of the stochastic equations (8.81) and (8.82).

In Figure 8.2 we show the result of a simulation where the parameters are $\lambda = 10$, $\mu = 0.1$, and $\nu = 10$. The initial conditions were chosen to be $M = 100$ and $J = 20$. This shows oscillations, in which the rises and falls of the population of the predators follows behind (lags) the rises and falls of the population of the prey. These oscillations are induced because the rise in the predator population induces a fall in the prey population. This reduction causes the predator population to fall because of the decrease in their food supply. This fall then allows the prey

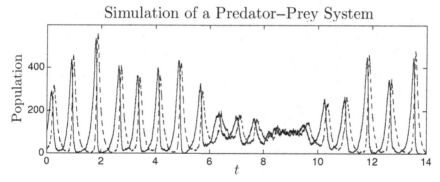

Figure 8.2. Population of the predator (dashed line) and the prey (solid line) as a function of time in a simulation of a simple stochastic model of a predator–prey system.

population to rise, which allows the predator population to rise, and the cycle continues. The fundamental cause of the oscillations is that the predator population always overshoots its available food supply, rather than reaching a steady-state.

We find that the size (amplitude) of the oscillations fluctuates. This is caused by the randomness in the jump events. If more prey are born than the average in some time period, and fewer happen to be eaten in the same time period, purely owing to chance, then the prey population will increase, followed by an increase in the predator population, leading to larger peaks in the oscillations.

It is worth noting that insight into the behavior of a jump process can be obtained by analyzing deterministic equations that give the approximate behavior of the means of the processes (in our case the means of M and J). These equations are obtained by making M and J real (continuous) deterministic variables, and making their rates of change equal to the average rates of change implied by the jump processes. From Eqs. (8.83)–(8.85), the deterministic equations for M and J are

$$\dot{M} = \lambda M - \mu J M, \tag{8.89}$$

$$\dot{J} = \mu J M - \nu J. \tag{8.90}$$

These are called the Lotka–Volterra equations. We simulate them with the same initial conditions and parameters as above, and the result is shown in Figure 8.3. We find that the populations oscillate in the same way as they do for their stochastic counterparts. In this case, however, the oscillations are exactly periodic, with a constant amplitude and wavelength. This confirms that the changes in the amplitude (and wavelength) in the stochastic predator–prey system are the result of the randomness.

An example of an interesting question one might wish to ask is the following: given some initial populations of predators and prey, what is the probability that

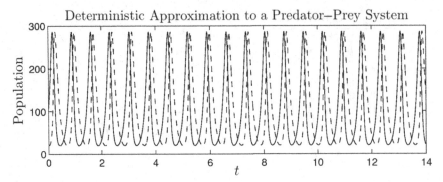

Figure 8.3. Population of the predator (dashed line) and the prey (solid line) as a function of time in a simple deterministic model of a predator–prey system.

they will go extinct? Such an event is only possible because of the randomness of the birth/death processes, and one expects it to be very small when both populations are large, and increase as they get smaller. One can answer this question, at least for the simple model above, by solving the master equation using a numerical simulation, and looking at the values of $P(j = 0, t) = \sum_m P(0, m, t)$ and $P(m = 0, t) = \sum_j P(j, 0, t)$ as time goes by. One can also get this information by calculating many realizations of the stochastic equations, and then averaging over them to obtain estimates of $P(n, m, t)$, and thus $P(j = 0, t)$ and $P(m = 0, t)$.

In fact, examples of the simple predator–prey model we have described are seldom seen in nature – this is because interactions between predators and prey usually involve many different species, not just two. It is seldom the case that a predator has a single food source, or that a given population has only a single predator. An exception to this is the case of Canadian lynxes and snowshoe hares, and data for the populations of these two species do indeed oscillate like the simple model. A discussion of the data on Canadian lynxes and hares may be found in *Fundamentals of Ecology* by Eugene Odum [29].

8.9 Biology: neurons and stochastic resonance

Neurons are cells that transmit signals to one another. The basic mechanism for this transmission is as follows. Each neuron has a number of inputs, these being ion currents that flow into them via a number of ion channels. The input current that flows in through each ion channel is controlled by other neurons, a different neuron controlling each channel. The ion currents come in two kinds, potassium and sodium. The sodium currents make a positive contribution to the total current flowing into a neuron, and the potassium currents make an effectively negative contribution (we won't bother to go into how this is achieved here). Once the total input current to a neuron reaches some threshold, the neuron changes its voltage with respect to its surroundings in an area near the ion channels. This localized voltage spike (called an "action potential") travels in a pulse down the rest of the neuron, which is a long thin tube called the *axon*. The end of the axon is the "output" of the neuron, and splits into many branches that connect to the inputs of other neurons. When the voltage pulse reaches the ends of the axon's branches, it causes an input current to flow into the inputs of these neurons.

In summary, the transmission of a signal from one neuron to another is accomplished when the first neuron changes its voltage for a short time, so as to generate an input to the second neuron. This second neuron will have many inputs from different neurons, and if enough of these inputs are stimulated, this neuron will in turn produce a voltage spike and in turn send a signal to the neurons to which its output is connected. The output signal of each neuron can be regarded as the voltage, V, at its outputs, and the inputs to a neuron as the ion currents at its input ion channels.

We now turn to the signals produced by neurons that transmit information to the brain from the ear. If the ear is exposed to a single note at a given frequency, f, then the neurons connected to the receptors in the ear produce a train of voltage spikes that are, on average, separated by time intervals of $1/f$. It appears from this that information coming from the ear about sound is encoded in the time between successive voltage spikes. The timing of the voltage spikes is noisy (that is, subject to random fluctuations), and it is the *average* time between successive spikes that appears to encode the frequency information. In order to obtain a simple model of an auditory neuron, we thus seek a noisy dynamical system that exhibits jumps between two states, and in which the average time between jumps is determined by the frequency of a sinusoidal signal.

It turns out that a simple model which satisfies the above requirements exhibits a surprising phenomena: adding noise to the system can increase the clarity of the output signal. This is referred to as *stochastic resonance*. In the neuron model, this means that adding some noise to the input can make the time intervals between successive output spikes fluctuate *less* about its average. The result of this is that, for each input frequency, there is a certain amount of input noise that minimizes the fluctuations of the spike train.

To motivate the model it helps to know the physical system from which it comes. Imagine a single particle trapped in one of two potential wells, separated by a potential barrier. If the particle is being kicked around by a noisy force, every now and then the particle will, by chance, get enough kinetic energy to hop over the barrier to the other well. The average rate at which this hopping occurs increases with the amount of noise. Now imagine that we apply a sinusoidal force, $F(t) = F_0 \cos(\omega t)$, to the particle. When the cosine is positive the force tries to push the particle from the left well to the right well, and when the cosine is negative it tries to push it the other way. When we choose F_0 so that the force is not strong enough to push the particle over the barrier on its own, it still influences the random hopping induced by the noise. The reason for this is that, during the half-period of the cosine in which the force is trying to push the particle over the barrier, the forces make it *easier* for the noise to cause the particle to cross the barrier, and thus increases the likelihood of it doing so. In this way the force effectively modulates the hopping rate. It is in this situation where there is a non-zero optimal noise level that will produce the most regular hopping between states (for a given sinusoidal force).

We could analyze the "double-well" system exactly as it is described above – by using the dynamical equations for the position and momentum of the particle in the double-well, and allow it to be kicked around by a force proportional to the Wiener noise, exactly as in the case of the Brownian particle in Chapter 5. However, there is a simpler model we can use instead that correctly reproduces the results of such an analysis. This builds in a simple way upon the model used to describe telegraph noise, discussed above in Section 8.5. In this model we include only two states, corresponding to the particle being in the left or right well, or equivalently the high and low voltage states of the neuron. We will label the low state as $N = 0$, and the high (spiking) state as $N = 1$. For $n = 0, 1$, we then allow the neuron to jump from $N = n$ to $N = 1 - n$ with random jump rate μ_n. Each of the two jump rates is some function of the noise strength and the sinusoidal force, F. Note that during the periods where the force increases the jump rate μ_n it should decrease the jump rate μ_{n-1}, which is in turn a fixed function of time. We choose the rates to be

$$\mu_1(t) = \mu e^{-r \cos(\omega t)}, \tag{8.91}$$

$$\mu_0(t) = \mu e^{r \cos(\omega t)}, \tag{8.92}$$

where ω is the frequency of the force, μ determines the overall average jump rate, and r determines the amount by which the force (being the input signal) affects the jump rates. The average jump rate in the absence of the force depends, of course, on the amount of noise. Denoting the strength of the noise by β, the relationship between the jump rate and the noise in the double-well system is

$$\mu(\beta) = \lambda e^{-k/\beta}, \tag{8.93}$$

where λ and k are constants determined by the shape of the wells. Finally, to correctly reproduce the results of the physical double-well system, we must set $r = r_0/\beta$. This means that the stronger the noise, the less influence the input signal has on the jump rates. We can also write these equations more compactly as $\mu_n = \mu e^{(-1)^n r \cos(\omega t)}$. Note that this model is merely the random telegraph process with time-dependent jump rates.

The dynamics of the neuron is given by the master equation

$$\dot{P}_0(t) = -\mu_1(t)P_0(t) + \mu_0(t)P_1(t), \tag{8.94}$$

$$\dot{P}_1(t) = -\mu_0(t)P_1(t) + \mu_1(t)P_0(t), \tag{8.95}$$

where $P_n(t)$ is the probability that $N = n$. Where convenient we will suppress the time argument, so that $P_n \equiv P_n(t)$. We will also use the alternative notation $P(n, t)$ for $P_n(t)$. We can use the fact that $P_0 + P_1 = 1$ to write an equation for P_1 alone, which is

$$\dot{P}_1 = -\gamma(t)P_1 + \mu_1(t) = -2r \cosh[\cos(\omega t)]P_1 + \mu e^{r\cos(\omega t)}, \tag{8.96}$$

where we have defined $\gamma(t) = \mu_0(t) + \mu_1(t)$. The solution to this differential equation is given in Section 2.4.1, and is

$$P_1(t) = e^{-\Gamma(t)}P_1(0) + \int_0^t e^{\Gamma(t')-\Gamma(t)}\mu_1(t')dt', \tag{8.97}$$

where

$$\Gamma(t) = \int_0^t \gamma(t')dt'. \tag{8.98}$$

We now assume that the parameter r is much less than unity ($r \ll 1$). This means that we can obtain approximate expressions for the transition rates by expanding the exponentials in Eqs. (8.91) and (8.92) in a Taylor series, and taking this series to second-order. This gives

$$\mu_n(t) \approx \mu \left[1 + (-1)^n r \cos(\omega t) + \frac{1}{2}r^2 \cos^2(\omega t)\right] \tag{8.99}$$

and

$$\gamma(t) \approx \mu \left[2 + r^2 \cos^2(\omega t)\right]. \tag{8.100}$$

With this approximation, the solution to the master equation is

$$P(n, t|n_0, 0) = \frac{1}{2}\left(1 + (-1)^n \kappa(t) + [2\delta_{n,n_0} - 1 - (-1)^n \kappa(0)]e^{-2\mu t}\right), \tag{8.101}$$

where n_0 is the state of the system at time $t = 0$, and

$$\kappa(t) = \left(\frac{2\mu r}{\sqrt{4\mu^2 + \omega^2}}\right)\cos(\omega t + \phi), \tag{8.102}$$

$$\phi = \arctan\left(\frac{\omega}{2\mu}\right). \tag{8.103}$$

What we really want to know is the strength of the periodic signal that appears at the output of the neuron (that is, in the flipping of the voltage between the two states)? One way to evaluate this is to calculate the mean value of the voltage as a function of time. If this oscillates sinusoidally at frequency f, then we know that the output does contain information about the input frequency. We can calculate $\langle N(t) \rangle$ directly from $P_1(t)$, and this gives

$$\langle N(t) \rangle = \sum_n P(n, t | n_0, 0) = P(1, t | n_0, 0)$$

$$= \frac{1}{2} \left(1 + \kappa(t) - [(-1)^{n_0} + \kappa(0)] e^{-2\mu t} \right). \tag{8.104}$$

As $t \to \infty$ the initial condition becomes unimportant, and we have

$$\langle N(t) \rangle = \frac{1}{2} + \frac{\mu r \cos(\omega t + \phi)}{\sqrt{4\mu^2 + \omega^2}}. \tag{8.105}$$

Indeed $\langle N(t) \rangle$ oscillates at frequency f, and thus the output signal contains information about the input frequency. The important quantity is the *amplitude* of the oscillation of $\langle N(t) \rangle$, as this indicates the clarity of the output signal regarding the value of f. In particular we would like to know how this amplitude depends on the noise strength μ. The amplitude of the oscillation in $\langle N(t) \rangle$ is

$$A(\mu) = \left(\frac{\mu}{\sqrt{4\mu^2 + \omega^2}} \right) r = \left(\frac{\mu(\beta)}{\beta \sqrt{4\mu(\beta)^2 + \omega^2}} \right) r_0, \tag{8.106}$$

where $\mu(\beta)$ is given by Eq. (8.93). It is now simple to show that A reaches a maximum at a value of β greater than zero. First we note that when $\beta = 0$, $A = 0$. This becomes obvious by making the replacement $y = 1/\beta$:

$$\lim_{\beta \to 0} A(\beta) = \lim_{y \to \infty} \left(\frac{y}{e^{ky} \sqrt{4e^{-2ky} + (\omega/\lambda)^2}} \right) r_0 = \lim_{y \to \infty} \left(\frac{\omega r_0}{\lambda} \right) \left(\frac{y}{e^{ky}} \right) = 0. \tag{8.107}$$

Similarly, as $\beta \to \infty$, $A(\beta) \to 0$. Clearly, however, for $\beta \in (0, \infty)$, $A(\beta)$ is greater than zero, and thus must reach a maximum for some value of β greater than zero. This is the phenomena of stochastic resonance. When the noise is at its optimal value, the time intervals between consecutive flips of the neuron's voltage, and thus between the voltage spikes, fluctuate less about their mean value. This produces the clearest signal at the output.

Further reading

Further discussion of predator–prey systems, including experimental data on populations of lynxes and hares in Canada, can be found in *Fundamentals of Ecology* by Eugene Odum [29]. Many more details regarding the theory of stochastic resonance, including applications to neurons, may be found in the review article by Gammaitoni *et al.* [30].

Exercises

1. For the Poisson process, calculate the expectation values of N^3 and e^N.
2. Determine the probability density for the time, τ, at which the first jump of the Poisson process occurs.
3. Consider a jump process in which N starts with the value 0, and jumps by two with the average rate λ. Solve the master equation for this process to determine $P(n, t)$.
4. Consider a stochastic process $N(t)$ that can take the values $N = 0, 1, 2, \ldots, \infty$, and which undergoes jumps of size $+1$ and -1. The rates of both processes are functions of N: the positive (or upward) jumps occur at the rate $\gamma_+(N)$, and the negative (downward) jumps occur at the rate $\gamma_-(N)$. In this case the steady-state set of probabilities, $P_{ss}(n) \equiv$ Prob$(N = n)$, can be obtained by choosing $P_{ss}(n)$ so that the total upward flow of probability from each state n to $n + 1$ matches the downward flow from $n + 1$ to n. This condition is called *detailed balance*. (More generally, detailed balance implies that the flow of probability between any two states is matched by an opposite flow between the same states.) Show that the steady-state probability for state $N = n$ implied by detailed balance is

$$P_{ss}(n) = \prod_{m=1}^{n} \frac{\gamma_+(m - 1)}{\gamma_-(m)} P_{ss}(0), \tag{8.108}$$

with $P_{ss}(0)$ chosen so that $\sum_{n=0}^{\infty} P_{ss}(n) = 1$.
5. Using the generating function method to solve the master equation for the Poisson process, calculate the generating function for the Poisson process when the initial condition is $N = k$. Then use the generating function to calculate the mean and variance for the Poisson process with this initial condition.
6. Consider the Poisson process, but now where the jump rate λ is some fixed function of time, so that $\lambda = f(t)$. What is $P(n, t)$ in this case?
7. Calculate the mean and variance of the telegraph noise process.
8. Solve the master equation for the random telegraph process when the transition rates are $\mu_0 = 1 - \cos(\omega t)$ and $\mu_1 = 1 + \cos(\omega t)$. Calculate the mean as a function of time.
9. There is a little round, fluffy, alien creature that is a popular pet on the planet Zorg. The color of the creatures' fur can be either red or blue, and will change spontaneously from one to the other at apparently random times. However, the creatures are social animals and this causes them to change color faster the more creatures there are of the other color. So if there are a number of the little guys in a box, the rate at which each one switches color is a constant, plus a term proportional to the number of creatures of the color it is switching to. Given that the total number of creatures in the box is N, write down the master equation for the probability that M of them are blue. What is the master equation when there are only two creatures in the box? Write this master

equation as a linear differential equation for the vector $\mathbf{x} = (x, y)^{\mathrm{T}}$, where $x \equiv P(0, t)$ and $y \equiv P(2, t)$.

10. In Chapter 7 we described chemical reactions using Fokker–Planck equations, but master equations can also be used to model chemical reactions. Consider a chemical reaction in which there are two reagents, A and B, in solution in a beaker. Let us denote the number of molecules of the first substance by N_A and the second by N_B. The chemicals react to produce a third substance, C, whenever a molecule of A collides with a molecule of B. If the rate at which this happens is $r A B$, write down the master equation for the three quantities N_A, N_B and N_C.

11. Consider a chemical reaction in which there are two reagents, A and B, as above. However, this time two molecules of A and one of B are required for the reaction to occur, in which case one molecule of C and also one molecule of A is produced. In this case, the rate at which the reaction occurs is $r A(A - 1)B$ (why?). Write down the master equation for the three quantities N_A, N_B and N_C in this case.

12. Write down the master equation for the chemical reaction

$$2A \rightarrow B$$
$$B \rightarrow 2A$$

where the two reactions occur at different rates. What is the master equation when there are only two molecules of A to start with? Solve the master equation in this case, and calculate the mean of A in the steady-state.

9

Levy processes

9.1 Introduction

So far we have studied stochastic differential equations driven by two fundamentally different noise processes, the Wiener process and the Poisson process. The sample paths of the Wiener process are continuous, while those of the Poisson process are discontinuous. The sample paths of a stochastic process $x(t)$ are continuous if its increments, dx, are infinitesimal, meaning that $dx \to 0$ as $dt \to 0$. The Wiener and Poisson processes have two properties in common. The first is that the probability densities for their respective increments do not change with time, and the second is that their increments at any given time are independent of their increments at all other times. The increments of both processes are thus mutually independent and identically distributed, or i.i.d. for short. In Section 3.3, we discussed why natural noise processes that approximate continuous i.i.d. processes are usually Gaussian, and that this is the result of the central limit theorem. In this chapter we consider *all* possible i.i.d. noise processes. These are the *Levy processes*, and include not only the Gaussian and Poisson (jump) processes that we have studied so far, but processes with continuous sample paths that do not obey the central limit theorem.

There are three conditions that define the class of Levy processes. As mentioned above, the infinitesimal increments, dL, for a given Levy process, $L(t)$, are all mutually independent and all have the same probability density. The third condition is a little more technical: the increment of a Levy process in a time step Δt, which we denote by ΔL, must satisfy

$$\lim_{\Delta t \to 0} \text{Prob}(\Delta L > \varepsilon) = 0, \tag{9.1}$$

for all values of ε greater than zero. Here the term $\text{Prob}(\Delta L > \varepsilon)$ is to be read as "the probability that ΔL is greater than ε". This condition is referred to as "stochastic continuity", and is quite reasonable. While this condition allows the infinitesimal increments dL to be finite (jump processes), it states that their probability of being finite goes to zero as dL goes to zero. The Poisson process obeys this, because the probability that the increment dN is unity is λdt, and thus goes to zero as $dt \to 0$. For the same reason, this condition also removes the possibility that there are jumps at fixed (deterministic) times: a jump at a fixed time, t, has a non-zero probability of occurring at that time. What stochastic continuity does not do is to prevent *infinitesimal* increments from being deterministic. The deterministic increment dt is a perfectly valid increment for a Levy process. This means that a Levy process can always have a deterministic component that increases linearly with time. This is referred to as a *drift*.

The two processes that we have studied so far, the Wiener process and the Poisson process, are Levy processes. But there are also more exotic Levy processes. We first describe an interesting special class of these exotic processes, called the *stable* Levy processes.

9.2 The stable Levy processes

We recall now that the Gaussian probability density is special, in that if we add two independent Gaussian random variables together, the result is also Gaussian. If the means of the two initial variables are zero, then the mean of the sum is also zero. The only difference between the density of the initial variables, and that of their sum, is that the standard deviation of the sum is $\sigma_{\text{sum}} = \sqrt{\sigma_1^2 + \sigma_2^2}$, where σ_1 and σ_2 are the standard deviations of the original variables. If $\sigma_1 = \sigma_2$, then the standard deviation of the sum is simply $\sqrt{2}$ times that of the original variables. If we add N i.i.d. Gaussian variables together, then the result has a standard deviation that is $\sqrt{N}$ times that of the original variables. Since the standard deviation is a measure of the *width* of the density, increasing the standard deviation by a factor of $\sqrt{N}$ is the same as stretching the density by the same factor. Note that to stretch a function, $P(x)$, by the factor a, we replace x by x/a; $P(x/a)$ is the stretched version of $P(x)$.

Recall from Chapter 1 that the probability density of the sum of two random variables is the convolution of their respective densities. Thus when we convolve two Gaussians together, the result is also a Gaussian. The central limit theorem tells us that when we convolve two densities with finite variances, the result becomes closer to a Gaussian. Now we ask, can we find densities that, when convolved with another copy of themselves, are not Gaussian but nevertheless retain their form apart from a simple stretching? Such densities would not obey the central limit theorem, and would therefore have infinite variances. We can find such densities by focussing first on their characteristic functions. Convolving a density with itself is the same as squaring its characteristic function. Further, scaling a density by a factor a is the same as scaling its characteristic function by $1/a$. Thus all we need to do is find functions that have the property that their square is a scaled version of themselves. The following exponential functions have this property: if we define $\chi_\alpha(s) = \exp\{-c|s|^\alpha\}$ (where c and α are non-negative constants), then

$$\chi_\alpha^2(s) = \exp\{-2c|s|^\alpha\} = \exp\{-c|2^{1/\alpha}s|^\alpha\} = \chi_\alpha(2^{1/\alpha}s). \tag{9.2}$$

Thus squaring $\chi_\alpha(s)$ *squashes* it by a factor $2^{-1/\alpha}$. Adding two random variables whose density has the characteristic function $\chi_\alpha(s)$, therefore results in a random variable with the same density, but *stretched* by the factor $2^{1/\alpha}$. The Gaussian density is a special case of this, since its characteristic function is $\chi_2(s) = \exp\{-\sigma^2 s^2\}$, giving the stretching factor $2^{1/2}$ (here σ is, as usual, the standard deviation of the Gaussian). The probability densities that correspond to these characteristic functions are called *stable* densities, because their form does not change under convolution.

If we write the constant c as $c = \sigma^\alpha$, then σ is a measure of the width of the corresponding probability density. (This is because multiplying σ by some factor k stretches the probability density by k.) It is therefore usual to write the characteristic functions as

$$\chi_\alpha^2(s) = \exp\{-\sigma^\alpha|s|^\alpha\}. \tag{9.3}$$

So what probability densities do these characteristic functions correspond to when $\alpha \neq 2$? It turns out that α must be in the range $0 < \alpha \leq 2$ for the characteristic function to correspond to a normalizable probability density. There are only three values of α for which the corresponding probability density has a closed form. The first is, of course, the Gaussian. The second is the *Cauchy density*,

$$C_\sigma(x) = \frac{\sigma}{\pi[x^2 + \sigma^2]}, \tag{9.4}$$

corresponding to $\alpha = 1$, and the third is the *Levy density*,

$$L_\sigma(x) = \begin{cases} \sqrt{\sigma/(2\pi x^3)}\, e^{-\sigma/(2x)}, & x > 0 \\ 0, & x \leq 0 \end{cases} \tag{9.5}$$

corresponding to $\alpha = 1/2$. The Cauchy and Levy densities with $\sigma = 1$ are plotted in Figure 9.1, along with the Gaussian with $\sigma = 1$ for comparison. We see that the "wings" or "tails" of the Cauchy and Levy densities are much wider than those of the Gaussian.

The mean of the Cauchy density is finite (it is zero for the density in Eq. (9.4), but can be changed to anything else merely by shifting the density along the x-axis), while its variance is infinite. The Levy density is even more pathological, since even its mean is infinite! More generally, the means of the stable densities are finite when $1 \leq \alpha \leq 2$, and infinite when $\alpha < 1$. The *only* stable density that has a finite

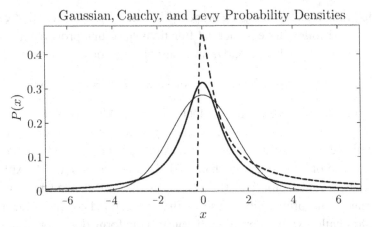

Figure 9.1. Here the Cauchy (thick solid line) and Levy (thick dashed line) are shown in comparison to the Gaussian thin solid line . For all the densities the width parameter $\sigma = 1$. (For the Gaussian σ is the standard deviation.) The means of the Cauchy and Gaussian densities are zero, and the peak of the Levy density has been chosen to coincide with those of the others.

variance is the Gaussian. The smaller α the more pathological the density. The stable densities are also referred to as the "α-stable" densities, to explicitly refer to the parameter that determines how they stretch when their random variables are added together.

Note that since the variance of the Cauchy density is infinite, the parameter σ that appears in Eq. (9.4) is not the standard deviation. It does, however, have a clear meaning in terms of the average value of the fluctuations of X from its mean value. Specifically, it is the average of the *absolute value* of the deviation of the variable from its mean:

$$\sigma \equiv \int_{-\infty}^{\infty} |x - \langle X \rangle| C(x) dx. \tag{9.6}$$

For the Levy density, however, the parameter σ cannot be related directly to a simple measure of the average fluctuations about the mean, since the mean is infinite.

The whole set of stable densities for $\alpha < 2$ now give Levy processes that are not Gaussian. This is achieved by choosing the increment of a Levy process to have an α-stable density. The resulting Levy processes are called the α-stable Levy processes, and we will denote their infinitesimal random increments by dS_α. To define the α-stable processes, we do, however, need to work out how the width of each stable density, σ_α, must scale with the time-step Δt. Recall from Chapter 3 that for a Gaussian the width (standard deviation), σ, of the increment, ΔW, has to be proportional to $\sqrt{\Delta t}$. The reason for this is that the sum of the increments for the two half intervals, $\Delta t/2$, must give the correct width for the increment for the full time interval when they are *added together*. If we denote the width of the Gaussian increment for time interval τ as $\sigma(\tau)$, then for the Gaussian we must have

$$\sqrt{2}\sigma(\Delta t/2) = \sigma(\Delta t) \tag{9.7}$$

and thus $\sigma(\Delta t) = \sqrt{\Delta t}$. For the increments of an α-stable density, we must have

$$2^{1/\alpha}\sigma_\alpha(\Delta t/2) = \sigma_\alpha(\Delta t), \tag{9.8}$$

and thus

$$\sigma_\alpha(\Delta t) = (\Delta t)^{1/\alpha}. \tag{9.9}$$

Let us consider as an example the Levy process driven by Cauchy increments. This is the 1-stable Levy process. Its infinitesimal increment, dS_1, is the Cauchy density

$$P(dS_1) = \frac{dt}{\pi\left[(dS_1)^2 + (dt)^2)\right]}, \tag{9.10}$$

because in this case $\sigma \propto dt$.

From our analysis above, we now know that the integral of the 1-stable, or Cauchy, process over a time T,

$$S_1(T) = \int_0^T dS_1(t), \tag{9.11}$$

has a Cauchy probability density (because the process is stable under summation, just like the Gaussian), and its width is given by $\sigma = T$. For a general α-stable process, its integral, $S_\alpha(T)$, has a width $T^{1/\alpha}$. Thus the smaller α, the more rapidly the width of the integrated process increases over time.

Before we move on, we present the complete set of characteristic functions for the densities that generate all the stable Levy processes, as these are somewhat more general than those we have discussed above. The complete set of stable Levy processes is generated by increments with the characteristic functions

$$\chi_{(\alpha,\beta)}(s) = \begin{cases} \exp\left\{-\sigma^\alpha|s|^\alpha\left[1 - i\beta\,\text{sgn}(s)\tan(\pi\alpha/2)\right] + i\mu s\right\}, & \alpha \neq 1 \\ \exp\left\{-\sigma|s|\left[1 + i(2\beta/\pi)\,\text{sgn}(s)\ln(|s|)\right] + i\mu s\right\}, & \alpha = 1. \end{cases} \tag{9.12}$$

Here σ is once again the width (or scaling) parameter, and μ is a parameter that shifts the density. For those densities for the which the mean is well-defined ($\alpha \geq 1$), μ is the mean. The new parameter β is an asymmetry, or *skewness*, parameter. When $\beta = 0$, the densities for which the mean is finite are symmetric about $x = \mu$. When β is not equal to zero, then it causes these densities to become asymmetric. The

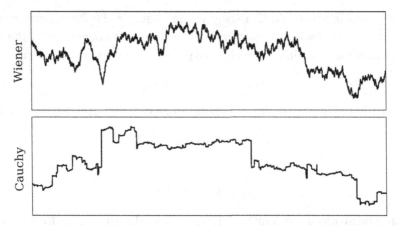

Figure 9.2. Sample paths of the Wiener and Cauchy processes. The width of the Wiener process scales as $\sqrt{t}$, and that of the Cauchy process as t.

value of β must be in the range $-1 \leq \beta \leq 1$. We still refer to all these Levy processes as being α-stable, regardless of the value of β, since β only affects the shape of the densities, not the fact that they are stable, nor the way they scale under convolution.

The sample paths of the stable Levy processes for $\alpha < 2$ can look quite different to those of the Wiener process. In Figure 9.2 we plot sample paths of the Cauchy process and the Wiener process for comparison.

9.2.1 Stochastic equations with the stable processes

Now that we have a bunch of new Levy processes, we can define differential equations that are driven by these processes:

$$dx = f(x,t)dt + g(x,t)dS_\alpha. \tag{9.13}$$

Because these processes are stable, just like Wiener noise, we can just as easily solve this stochastic equation for the case in which g is constant and f is linear in x (this is the stable Levy process equivalent of the Ornstein–Uhlenbeck equation):

$$dx = -\gamma x dt + g dS_\alpha. \tag{9.14}$$

The solution is

$$x(t) = x_0 e^{-\gamma t} + g \int_0^t e^{\gamma(s-t)} \, dS_\alpha(s). \tag{9.15}$$

The probability density for a random variable

$$Y(t) = g \int_0^t f(s) \, dS_\alpha(s), \tag{9.16}$$

for any function of time, $f(t)$, is that given by the relevant characteristic function in Eq. (9.12), with $\mu = 0$ and

$$\sigma = g \left[\int_0^t [f(s)]^\alpha ds \right]^{1/\alpha} . \tag{9.17}$$

This can be shown in the same way as for the Wiener process (Chapter 3), and we leave it as an exercise.

The solution for the multivariate Ornstein–Uhlenbeck process, in which the Wiener noises are replaced by α-stable processes, is also just the same as that for Wiener noise given in Eq. (3.104), with the vector of noises $\mathbf{dW}$ replaced by a vector of α-stable noises $\mathbf{dS}_\alpha$.

Unfortunately things do not remain that simple for more complex stochastic equations. For even the full linear stochastic equation, given by

$$dx = -\gamma x dt + g x dS_\alpha, \tag{9.18}$$

no closed-form expression is known for the probability density (or characteristic function) of the solution. Recall that to solve this equation for Gaussian noise ($\alpha = 2$), we used the fact that $(dW)^2 = dt$, which gives

$$x + dx = x - \gamma x dt + g x dW = e^{-(\gamma + g^2/2)dt + g dW} x. \tag{9.19}$$

But we cannot make this transformation for any of the other stable processes, because

$$\langle (dS_\alpha)^2 \rangle = \infty \quad \text{for} \quad \alpha \neq 2. \tag{9.20}$$

Given this fact, one might even wonder whether the solution to the linear stochastic equation, Eq. (9.18), is well defined at all! One need not worry, however; one merely has to recall that random variables are well defined even if their variances, or even their means, are infinite. While there is no closed-form solution to Eq. (9.18), there is an explicit expression for the solution, which we show how to derive in Section 9.4.

9.2.2 Numerical simulation

One way to simulate stochastic equations driven by α-stable Levy processes, is to generate random variables whose densities are the α-stable densities. It turns out that while the densities for almost all the α-stable processes have no known closed form, there is a simple and efficient method for numerically generating samples from all of these densities. There are a number of algorithms to do this [31, 32]. We will first give some particularly simple formulae for some special cases, then give the formula of Chambers *et al.* for all symmetric cases ($\beta = 0$), and finally give Kanter's method for the general case [31]. We also note that the method of Chambers *et al.* does extend to the general case, and is more efficient that Kanter's method, but contains a subtlety when α is close to unity. This algorithm may be found in Ref. [32].

All modern programming languages include built-in functions to generate random numbers sampled from a uniform probability density on the interval [0, 1]. This can be transformed to a variable with a uniform density on any fixed interval by scaling and shifting. The algorithms we describe therefore take such random variables as their basic ingredients. In the following, $S_{(a,b)}$ will denote an α-stable random variable with $\alpha = a, \beta = b, \sigma = 1$ and $\mu = 0$. We will also need a random variable, y, with a "standard" exponential density. We therefore begin by describing how to generate this random variable.

The standard exponential density. This is the probability density

$$P(y) = \begin{cases} 0, & y < 0 \\ e^{-y}, & y \geq 0. \end{cases} \tag{9.21}$$

To generate a random variable, y, with this probability density we use the formula

$$y = -\ln x, \tag{9.22}$$

where x has a uniform density on [0, 1], and we exclude (throw away) the values $x = 0$ and $x = 1$.

The Cauchy density. A Cauchy random variable is given by

$$S_{(1,0)} = \tan(x), \tag{9.23}$$

where x has a uniform density on $[-\pi/2, \pi/2]$, and we exclude the endpoints $x = \pm\pi/2$.

The stable density with $\alpha = 1/2$ and $\beta = 1$. This is given by

$$S_{(1/2,1)} = 1/w^2, \tag{9.24}$$

where w is a zero mean Gaussian random variable with variance $V = 2$. Naturally in this expression we have to throw away the cases in which $w = 0$. (A method for generating Gaussian random variables is given in Section 6.1.1.)

All stable densities with $\beta = 0$. These are given by the formula [32]

$$S_{(\alpha,0)} = \frac{\sin(\alpha x)}{(\cos x)^{1/\alpha}} \left(\frac{\cos\left[(1-\alpha)x\right]}{y} \right)^{(1-\alpha)/\alpha}, \tag{9.25}$$

where x has a uniform density on the interval $[-\pi/2, \pi/2]$, and y has the standard exponential density described above.

Kanter's method for all stable densities with $\beta = 1$. These random variables are given by

$$S_{(\alpha,1)} = \frac{\sin(\alpha x)}{(\sin x)^{1/\alpha}} \left(\frac{\sin\left[(1-\alpha)x\right]}{y} \right)^{(1-\alpha)/\alpha}, \qquad (9.26)$$

where x has a uniform density on $[0, \pi]$, and y has the standard exponential density.

Kanter's method for all stable densities. Generate two independent stable random variables, A and B, each with the density of $S_{(\alpha,1)}$. We do this using Kanter's method above. The desired α-stable random variable is then given by

$$S_{(\alpha,\beta)} = \eta_+ A - \eta_- B, \qquad (9.27)$$

where

$$\eta_\pm = \left(\frac{\sin\left[(\pi/2)(1 \pm \beta)(1 - |1 - \alpha|)\right]}{\sin\left[\pi(1 - |1 - \alpha|)\right]} \right)^{1/\alpha}. \qquad (9.28)$$

9.3 Characterizing all the Levy processes

It turns out that there is a remarkably elegant way to characterize *all* possible Levy processes. This is because the characteristic function for every Levy process has a rather simple form. To introduce this form, and to understand the meaning of its various parts, we should first examine the characteristic functions for random variables generated by the Wiener and Poisson processes, with the addition of deterministic drift.

To being with, a Levy process with drift and a single Wiener noise is given by the stochastic equation

$$dz = \mu dt + \sigma dW, \qquad (9.29)$$

where μ is the drift rate, and σ is the size of the Wiener noise. The solution to this equation, $z(t)$, is a Gaussian random variable with mean μt, and variance $\sigma^2 t$. From Chapter 1 the characteristic function for the density of $z(t)$ is

$$\chi_z(s) = \exp\{it\mu s - t\sigma^2 s^2\}. \qquad (9.30)$$

So a term in the exponential that is linear in s describes a deterministic drift, and a term that is quadratic in s describes Gaussian noise.

We now turn to the Poisson process. Recall from Chapter 8 that the Poisson process, $N(t)$, with jump size δ can take the values $N(t) = \delta n$, for $n = 0, 1, 2, \ldots, \infty$. The set of probabilities for these values is

$$\text{Prob}(N(t) = \delta n) \equiv P(n, t) = \frac{e^{-\lambda t}(\lambda t)^n}{n!}. \qquad (9.31)$$

From this we can calculate the characteristic function, which is

$$\chi_{N(t)}(s) \equiv \sum_{n=0}^{\infty} e^{in\delta s} \left(\frac{e^{-\lambda t}(\lambda t)^n}{n!} \right) = \exp\{\lambda t(e^{i\delta s} - 1)\}. \tag{9.32}$$

Now consider a stochastic process $J(t)$ that is the sum of a number of Poisson processes all happening simultaneously, with different rates and different jump sizes. This is called a *compound* Poisson process. If we denote a Poisson process with rate λ and jump size δ by $N(\lambda, \delta, t)$, then $J(t) = \sum_{m=1}^{M} N(\lambda_m, \delta_m, t)$. Calculating the characteristic function for $J(t)$ is very simple, since the characteristic function for the sum of a number of random variables is merely the product of their respective characteristic functions. So the characteristic function for $J(t)$ is

$$\chi_{J(t)}(s) = \exp\left\{ t \sum_{m=1}^{M} (e^{i\delta_m s} - 1)\lambda_m \right\}. \tag{9.33}$$

So we see that the characteristic function for a compound Poisson process is obtained simply by adding together in the exponential the functions $(e^{i\delta_m s} - 1)\lambda_m$ for each Poisson process m.

We can now consider a more general compound Poisson process that has a continuum of different jump sizes, rather than a discrete set. The rate λ for having jumps of size δ is now a function $\lambda(\delta)$, which we will refer to as the "jump rate density". The characteristic function is now given by replacing the sum in the exponential with the integral of $\lambda(\delta)$ over δ:

$$\chi_{J(t)}(s) = \exp\left\{ t \int_{-\infty}^{\infty} (e^{i\delta s} - 1)\lambda(\delta)d\delta \right\}. \tag{9.34}$$

The total rate of jumps per unit time, λ, is the integral of the jump rate density over δ:

$$\lambda = \int_{-\infty}^{\infty} \lambda(\delta)d\delta. \tag{9.35}$$

If we divide $\lambda(\delta)$ by this total jump rate, the result is the probability density that the size of a jump, when it occurs, is equal to δ. Writing this probability density as $p(\delta)$, we can alternatively write the characteristic function as

$$\chi_{J(t)}(s) = \exp\left\{ \lambda t \int_{-\infty}^{\infty} (e^{i\delta s} - 1)p(\delta)d\delta \right\}. \tag{9.36}$$

Now let us examine the jump rate density $\lambda(\delta)$ in a bit more detail. Clearly this should integrate to a finite number for all jumps that are finite, otherwise the process $J(t)$ would instantly go to infinity. However, it turns out that we can have a perfectly well-defined Levy process in which the jump rate goes to infinity as the size of the jumps goes to zero. An infinite rate of infinitely small jumps is admissible. So let us arbitrarily break the integral over δ into two parts, one for $|\delta| > 1$, and one for $|\delta| \leq 1$. For the jumps that are larger than one the jump rate density, also called the *Levy measure*, must be finite:

$$\int_{|\delta|>1} \lambda(\delta)d\delta < \infty. \tag{9.37}$$

We said above that the jump rate can go to infinity as the size of the jumps goes to zero, and while this is true, it is very subtle. This causes the process to move beyond a compound jump process, and the distinction between jumps, drift and diffusion blurs. While the proofs of these subtleties are beyond us here, we can nevertheless provide some arguments for them. The first thing to note is that the function $e^{i\delta s} - 1$ in the exponential that describes jumps of size δ contains all powers of s:

$$e^{i\delta s} - 1 = i\delta s - \frac{1}{2}\delta^2 s^2 - i\frac{1}{3!}\delta^3 s^3 + \cdots . \tag{9.38}$$

We now recall from the start of this section that the first term in this power series expansion is precisely the same as the one that gives the process a deterministic drift. If we let $\lambda(\delta)$ tend to infinity as δ tends to zero, then the first term in the power series generates infinite drift. So before we allow λ to blow up at $\delta = 0$, we must remove this term in a finite region around $\delta = 0$. Our characteristic function then becomes

$$\chi_{J(t)}(s) = \exp\left\{ t \int_{|\delta|\leq 1} (e^{i\delta s} - 1 - i\delta s)\lambda(\delta)d\delta + t \int_{|\delta|>1} (e^{i\delta s} - 1)\lambda(\delta)d\delta \right\}. \tag{9.39}$$

It turns out that this characteristic function gives a well-defined Levy process even if λ blows up at $\delta = 0$, so long as

$$\int_{|\delta|\leq 1} \delta^2 \lambda(\delta)d\delta < \infty. \tag{9.40}$$

We can now present the remarkable fact that allows us to characterize all the Levy processes, a result referred to as the Levy–Khinchin representation: every Levy process has a characteristic function of the form

$$\chi_{J(t)}(s) = \exp\left\{ t\left[i\mu s - \sigma^2 s^2 \right] \right\}$$

$$\times \exp\left\{ t\left[\int_{|\delta|\leq 1} (e^{i\delta s} - 1 - i\delta s)\lambda(\delta)d\delta + \int_{|\delta|>1} (e^{i\delta s} - 1)\lambda(\delta)d\delta \right] \right\}, \tag{9.41}$$

where $\lambda(\delta)$ is called the *Levy measure* or *Levy density*, and satisfies Eqs. (9.37) and (9.40).

The above result means that every Levy process can be constructed as the sum of (1) a drift, (2) a Wiener process, (3) a compound Poisson process, and (4) a process that is specified as a jump process, but has an infinite rate of infinitesimal jumps. We can therefore divide Levy processes up into two groups – those that have a finite jump rate, and the exotic Levy processes that have an infinite rate. Those with an infinite jump rate are referred to as *infinite activity* Levy processes. All processes with an infinite variance are infinite activity.

The above result means, for example, that the Cauchy process can be simulated using jumps, instead of explicit Cauchy increments. The Levy densities for the α-stable processes have the form

$$\lambda(\delta) = \begin{cases} a\delta^{-(\alpha+1)}, & \delta > 0 \\ b|\delta|^{-(\alpha+1)}, & \delta < 0, \end{cases} \tag{9.42}$$

for some positive constants a and b; a discussion of this may be found on p. 94 of Ref. [33].

9.4 Stochastic calculus for Levy processes

For stochastic equations containing Levy processes we can derive a transformation formula that is the equivalent of Ito's formula (Eq. (3.40)) for Wiener noise. Deriving this formula for Levy processes with a finite activity (that is, in which the jumps are due only to a compound Poisson process) is simple, and it turns out that this formula, when written in the right form, is also true for Levy processes with infinite activity. (We have, in fact, already considered a simple version of this formula for jump processes in Chapter 8, namely the rule $dN^2 = dN$.)

To obtain the transformation formula for Levy calculus we consider a Levy process, $x(t)$, given by

$$dx = f(x, t)dt + g(x, t)dW + dJ, \tag{9.43}$$

where $J(t)$ is a compound Poisson process. We also define dx_c as the part of the increment of $x(t)$ that excludes the jumps, so that $dx_c = f(x, t)dt + g(x, t)dW$. Let us denote the times that the jumps occur as t_i, $i = 1, 2 \ldots, \infty$, and the sizes of the jumps as $\Delta J_i \equiv \Delta J(t_i)$.

We wish to work out the increment of a variable y, where y is a function of x and t. We know that the process consists of periods in which there are no jumps, so that the process is only driven by Wiener noise, punctuated by instantaneous jumps. First consider a period between the jumps. During this time the increments of $x(t)$ are equal to dx_c, so we can use Ito's formula:

$$dy = \left(\frac{\partial y}{\partial x}\right) dx_c(t) + \left(\frac{\partial y}{\partial t}\right) dt + \frac{g^2(x, t)}{2}\left(\frac{d^2 y}{dx^2}\right) dt. \tag{9.44}$$

Each time a jump occurs, the increment in y is merely

$$dy(t_i) = y(x_i + \Delta J_i) - y(x_i), \tag{9.45}$$

where we have defined x_i as the value of $x(t)$ just *before* the ith jump occurs. Since dy is the sum of these two, we have

$$dy(t) = \left(\frac{\partial y}{\partial x}\right) dx_c(t) + \left[\left(\frac{\partial y}{\partial t}\right) + \frac{g^2(x, t)}{2}\left(\frac{\partial^2 y}{\partial x^2}\right)\right] dt$$
$$+ [y(x + \Delta J(t), t) - y(x, t)],$$

$$\text{with} \quad \Delta J(t) = \begin{cases} \text{size of jump at } t_i, & t = t_i, \forall i \\ 0 & \text{otherwise} \end{cases}. \tag{9.46}$$

Writing this now in terms of $x(t)$ instead of $x_c(t)$ we have

$$dy(t) = \left(\frac{\partial y}{\partial x}\right) dx(t) + \left[\left(\frac{\partial y}{\partial t}\right) + \frac{g^2(x, t)}{2}\left(\frac{\partial^2 y}{\partial x^2}\right)\right] dt$$
$$+ \left[y(x + \Delta J(t), t) - y(x, t) - \frac{dy(x, t)}{dx}\Delta J(t)\right]. \tag{9.47}$$

In this expression the term $-\frac{dy}{dx}\Delta J(t)$, in the sum merely cancels the extra contribution generated by the first term when we replace x_c with x. While the formulae given by Eqs. (9.46) and (9.47) are the same for finite-activity Levy processes, it is only the second, Eq. (9.47), that gives the correct formula for infinite activity Levy processes [33].

9.4.1 The linear stochastic equation with a Levy process

Using the Levy calculus formula, Eq. (9.47), one can obtain an explicit expression for the solution to the linear stochastic equation

$$dx = -\gamma x dt + x dL, \tag{9.48}$$

where dL is the increment of any Levy process. To obtain the solution we make the now familiar transformation $y = \ln(x)$. But before we do this we write the Levy process in three parts:

$$dL = \beta dt + g dW + \Delta L_J. \tag{9.49}$$

The first two terms are the drift and Wiener noise components of the Levy process. The second term, ΔL_J, is the part that is described by jumps (see Eq. (9.41)). Recall that these jumps may generate probability densities with infinite variance, such as the Cauchy process. Now using the transformation rule given in Eq. (9.47), we have

$$dy = -\left(\gamma - \frac{g^2}{2}\right)dt + dL + \ln(x + \Delta x) - \ln(x) - \frac{\Delta x}{x},$$

$$= -\left(\gamma - \frac{g^2}{2}\right)dt + dL + \ln\left(1 + \frac{\Delta x}{x}\right) - \frac{\Delta x}{x}, \quad (9.50)$$

where Δx is the *discontinuous* part of the change in x (the "jump") at time t. These jumps come purely from the process ΔL_J. Now we note that

$$\Delta x = x\Delta L_J. \quad (9.51)$$

Substituting this into Eq. (9.50), we have

$$dy = -\left(\gamma - \frac{g^2}{2}\right)dt + dL + \ln(1 + \Delta L_J) - \Delta L_J, \quad (9.52)$$

and we can now just integrate both sides to obtain

$$y(t) = y_0 - \left(\gamma - \frac{g^2}{2}\right)t + L(t) + \sum_{n=1}^{N} \ln\left[1 + \Delta L_J(t_n)\right] - \Delta L_J(t_n). \quad (9.53)$$

Here the index n enumerates the times at which the jumps occur, and N is the total number of jumps in the interval $[0, t]$. The index n could be discrete, as for a compound Poisson process, or continuous. In the latter case N is no longer relevant, and the sum becomes an integral. Exponentiating the above equation to obtain $x = e^y$, the final solution is

$$x(t) = x_0 e^{-(\gamma - g^2/2)t + L(t)} \prod_{n=1}^{N} [1 + \Delta L_J(t_n)] e^{-\Delta L_J(t_n)}. \quad (9.54)$$

Further reading

Further details regarding Levy processes are given in the comprehensive introductory text *Financial Modelling with Jump Processes* by Rama Cont and Peter Tankov [33]. Applications of Levy processes to finance are also given there. Applications of Levy processes to a range of phenomena in the natural sciences may be found in the review article by Ralf Metzler and Joseph Klafter [34], and references therein.

Exercises

1. Calculate the characteristic function for the Poisson process.
2. By multiplying the relevant characteristic functions together, and taking the continuum limit, derive Eq. (9.17).

3. Consider a process $x(t)$ that is the sum of: (1) A Poisson process with jump rate λ; and (2) a deterministic drift with rate $-\lambda$. What is the mean of x as a function of time? (Note: the process x is called a *compensated* Poisson process.)

4. Determine the characteristic function of the process x defined by

$$dx = -\mu x dt + \beta(t)dW + dN,$$

where dN is the increment of a Poisson process with jump rate λ.

5. Solve the stochastic equation

$$dx = -\mu x dt + t dC,$$

where dC is the increment of the Cauchy process.

6. Use the solution given in Eq. (9.54) to write down the solution to the linear stochastic equation

$$dx = \alpha x dt + \beta x dN,$$

and show that it agrees with the solution obtained in Section 8.2.

7. Write down an explicit solution to the stochastic equation

$$dx = -\mu(t)x dt + g(t)dL,$$

where dL is the increment of a Levy process. Determine the probability densities of the terms that appear in the solution where possible.

10

Modern probability theory

10.1 Introduction

You are probably asking yourself why we need a second chapter on probability theory. The reason is that the modern formalism used by mathematicians to describe probability involves a number of concepts, predefined structures, and jargon that are not included in the simple approach to probability used by the majority of natural scientists, and the approach we have adopted here. This modern formalism is not required to understand probability theory. Further, in the author's experience, the vast majority of physically relevant questions can be answered, albeit non-rigorously, without the use of modern probability theory. Nevertheless, research work that is written in this modern language is not accessible unless you know the jargon. The modern formalism is used by mathematicians, some mathematical physicists, control theorists, and researchers who work in mathematical finance. Since work that is published in these fields is sometimes useful to physicists and other natural scientists, it can be worthwhile to know the jargon and the concepts that underly it.

Unfortunately a considerable investment of effort is required to learn modern probability theory in its technical detail: significant groundwork is required to define the concepts with the precision demanded by mathematicians. Here we present the concepts and jargon of modern probability theory *without* the rigorous mathematical technicalities. Knowing this jargon allows one to understand research articles that apply this formalism to problems in the natural sciences, control theory, and mathematical finance. Further, if the reader wishes to learn the mathematical details, being pre-equipped with these underlying concepts should smooth the task considerably. There are many good textbooks that present the details of the modern formalism; we recommend, for example, the excellent and concise text *Probability with Martingales* by David Williams [35]. We hope that this chapter will be especially useful to people working in interdisciplinary fields in which there

166

remains some division between research communities who use one or the other formalisms, and contribute to smoothing such divisions. In the following we have placed all the key jargon terms in bold italic, so that they are easy to locate for reference purposes. We have also included some additional technical information under the heading *Technical notes*. These can be skipped without affecting the rest of the presentation.

10.2 The set of all *samples*

The first thing we must define when considering probabilities, is the set of all mutually exclusive possible outcomes. One, and only one, of these individual outcomes *will* happen. If we roll a six-sided die, then there are six mutually exclusive possibilities for the number that will appear face-up on the die. The mutually exclusive possibilities are called the ***samples***. This term comes from the notion of "sampling" a probability density (when we look to see what the value of a random variable is, we are said to have *sampled* the random variable). The set of all the samples is denoted by the symbol Ω, and called the ***sample space***. When describing a die roll, the set Ω has six elements. An element of Ω (that is, a sample), is denoted by ω.

10.3 The collection of all *events*

Rather than asking precisely what sample has occurred when we roll a die, we could ask whether the number on the die is greater than three, or whether the number is even. Each of these conditions (the number being greater than three, or being even) has a probability of occurring. Each of these conditions will be true for *some* of the samples, but not for all of them. Thus each condition corresponds to a *subset* of the samples in the sample space. Each possible subset of the samples is called an ***event***. If we roll the die and obtain a sample that is greater than three, then we say that the event "the number on the die is greater than three" has occurred. The collection of all possible events is called the "σ-algebra" (pronounced "sigma-algebra") of events, for reasons we explain next.[1]

10.4 The collection of events forms a *sigma-algebra*

An algebraic structure, or *algebra*, is a collection of objects, along with a number of mathematical operations that can be performed on the objects. These operations can take a single object, in which case they produce a second object, and are referred to as *unary*. Operations may also take two objects, in which case they produce a

---

[1] We call the set of events a collection rather than a set because of Russell's paradox involving sets of sets [36].

third and are referred to as *binary*. To be an "algebra" the collection must be *closed* under the operations. This means that whenever an operation is performed on one or more of the objects, the object that it produces is also part of the collection.

Now consider the collection of all possible events for a given scenario (such as rolling a die). Since each event is a set (of the samples), there are two natural binary operations that can be performed on the events: *union* and *intersection*. The union of two sets is the set of all the elements that appear in *either* of the sets, and the intersection is the set of all elements that appear in *both* the sets. If we denote one event as A, and another as B, then the union of the two is denoted by $A \cup B$, and the intersection by $A \cap B$. There is also an operation that we can perform on a single event. This operation gives the *complement*, also called the *negation* of the event. The complement of an event is the set of all the elements in the sample space that are *not* contained in the event. The complement of the event A, also referred to as "not A", is denoted by $\neg A$. Thus the union of an event, A, and its complement, $\neg A$, is the whole sample space: $\Omega = A \cup \neg A$. Finally, we can also speak of the union (or intersection) of more than two events. Naturally the union of a number of events is the set of samples that are contained in *any* of the events, and the intersection is the set of samples that are contained in *all* of the events. These operations on more than two events can always be written in terms of the their binary counterparts. For example, the union of the three sets A, B, and C is given by $A \cup (B \cup C)$.

Because the collection of all the events is closed under the operations $\cup$, $\cap$, and $\neg$, it is an *algebra* (in particular it is called a Boolean algebra). If the number of samples is finite, then this is all there is to it. However, the number of samples is often infinite. For example, if one is measuring the length of an object, then the result of this measurement has an infinite number of possible values. Mathematicians have found that in this case, to ensure the answer to every question we might want to ask is well-defined, the collection of events must be closed under the union of a *countably infinite* number of events. By "countably infinite" we mean that the events in the union can be put in one-to-one correspondence with the positive integers. A collection of sets with this extra condition is called a **sigma-algebra**, or σ-algebra. The σ-algebra of all the events in the set Ω is usually denoted by $\mathcal{F}$ or Σ.

Technical note. The σ-algebra *generated* by a collection of sets is defined as the smallest σ-algebra that contains all the sets in the given collection. The most useful σ-algebra is the **Borel** σ-algebra of the real numbers. This is the σ-algebra that contains (is generated by) all the open intervals on the real line. A vector of N real numbers is a member of the N-dimensional space $\mathbb{R}^N$, and the N-dimensional Borel σ-algebra is the σ-algebra containing all the open N-dimensional cubes in $\mathbb{R}^N$. A set that is a member of a Borel σ-algebra is called a **Borel set**.

10.5 The probability *measure*

So far we have the set of all the samples, Ω, called the *sample space*, and the σ-algebra of all the events, which is denoted by $\mathcal{F}$. We need one more thing to complete the basic building blocks of our probability theory. We need to assign a probability to each of the samples, w. More generally, we need to assign a probability to every event. A rule, or map, that associates a number (in our case the probability) with every element of a σ-algebra is called a *measure*.

The reader will already be family with the everyday "measure" defined by integration. Consider a real variable x, where x can take any value on the real line. The set of all values of x is the sample space, Ω. Every segment of the real line, $[a, b]$, where $0 \leq a \leq b \leq 1$, is a subset of the sample space, and thus an event. The integral

$$\int_a^b dx = b - a \qquad (10.1)$$

associates a real number with each of these sub-intervals. Note that we can change the measure by introducing a function, $f(x)$, and associating the number $\int_a^b f(x)dx$ with each interval $[a, b]$. It is from integration that the notion of a measure first sprung. A primary reason for the development of measure theory is that there are many useful subsets of the real line that standard integration (Riemann integration) is not sufficient to handle. It will provide a number for every intersection of a finite number of intervals, but one has to develop a more sophisticated measure to consistently assign a number to every *countably infinite* intersection of intervals. It is the *Lebesgue* measure that does this. One can broadly think of the Lebesgue measure as associating a weighting with each point on the real line. We can then talk of integrating over a segment of the real line *with respect to this measure*.

If we define a measure, μ, that maps the interval $[a, b]$ to the number c, then we can write $c = \mu([a, b])$. Alternatively, since the Lebesgue measure (and, in fact, any measure) must have all the basic properties of the usual integral, it is natural to use an integral notation for a measure. This notation is

$$c = \int_a^b \mu(dx). \qquad (10.2)$$

More generally, if our sample space is not the real line, then we can write the number that the measure gives for an event A as

$$c = \int_A \mu(d\omega) = \int_A d\mu. \qquad (10.3)$$

This formula can be read as "c is the integral over the samples in the set A, using the measure μ".

A *probability measure*, $\mathbb{P}$, is a measure that gives a non-negative number for every event, and which gives unity for the whole sample space:

$$1 = \int_\Omega \mathbb{P}(d\omega). \tag{10.4}$$

The probability measure must give unity for the whole space, since one of the samples in the sample space *must* occur. (In the example of rolling the die, we know that one of the numbers will appear face up, we just don't know which one.) We also note that the σ-algebra of the events includes the empty set. This is the set that contains no samples at all, and is denoted by $\emptyset$. The probability for the empty set is zero, also because one of the samples must occur.

Integrating a function

A *function* on the sample space is a function that maps each element, ω, to a number. In addition to calculating the measure associated with a set of samples, we also need to be able to integrate a function over the space of samples. The integral of a function is defined with respect to a particular measure. This means, loosely speaking, that to integrate a function over a set A, we cut A up into tiny subsets, multiply the measure for each subset by a value that the function takes in that subset, and then add up the results for all the subsets. If the measure we use is the Lebesgue measure, then this corresponds to our usual definition of the integral of a function (the Lebesgue measure is the same as the Riemann integral for simple intervals). We write this as

$$\int_A f(\omega)\mu(d\omega) \quad \text{or} \quad \int_A f d\mu. \tag{10.5}$$

The Radon–Nikodým derivative

Given a probability measure $\mathbb{P}$, one can define a new probability measure $\mathbb{Q}$ by

$$\int_A f(\omega)\mathbb{Q}(d\omega) = \int_A f(\omega)g(\omega)\mathbb{P}(d\omega). \tag{10.6}$$

To obtain $\mathbb{Q}$ we are thus "multiplying" the measure $\mathbb{P}$ by the function g. This function is called the *Radon–Nikodým derivative* of $\mathbb{Q}$ with respect to $\mathbb{P}$, and is written as

$$\frac{d\mathbb{Q}}{d\mathbb{P}} = g. \tag{10.7}$$

One also speaks of $\mathbb{Q}$ as having *density g* with respect to $\mathbb{P}$.

Technical note. The **Lebesgue measure** on the real line is based on the Borel σ-algebra. To obtain the Lebesgue measure one first defines a simple measure μ_0

that maps every open interval (a, b) to the value $b - a$. The set of open intervals generates the Borel σ-algebra. This means that all the sets in the Borel σ-algebra are obtained by taking countably infinite unions and intersections of the open intervals. The corresponding countably infinite additions and subtractions of the values $b - a$ then give the values for all the sets in the Borel σ-algebra. This is the Lebesgue measure, which assigns a value to every Borel set.

10.6 Collecting the concepts: random variables

Now we have everything we need to describe a random outcome: we have a set of possible samples, and can assign a probability to every event. The three key structures are the sample space, Ω, the sigma-algebra of events, $\mathcal{F}$, and the probability measure, $\mathbb{P}$. Together these three structures are called the ***probability space***. The probability space is usually denoted by the list $(\Omega, \mathcal{F}, \mathbb{P})$.

Random variables

In the simple description of probability presented in Chapter 1, the main objects are random variables and probability densities. So now we want to see how these are defined in terms of the new structures we have described above. A random variable is an object that can take a range of possible values. So far we have defined the possible outcomes, ω, but we have not given these outcomes (samples) any particular *values* – the samples are completely abstract. We now define a random variable as something that has a value for each of the samples, ω. Thus when a given outcome occurs, the random variable takes a definite value. Note that because the random variable maps each of the samples ω to a value, it is actually a *function* from the sample space Ω, to a set of values. Often this set is the real numbers (for example a Gaussian random variable) or is a finite set of integers (the result of a die roll).

Every subset of the possible values of the random variable corresponds to a subset of the samples, and thus to an event. The probability measure gives us a probability for every subset of the values that the random variable can take. This is what the probability density does for us in the simple formalism.

Probability densities

It is not necessary for our purposes to know precisely how the probability density is related to the probability measure. If you do want to know, we can describe the relationship with the following simple example. Let us say that we have a random variable X that takes values on the real line, and is defined by the function $f(\omega)$. Recall that the measure on the probability space is $\mathbb{P}$. The probability that $-\infty < X < x$ (recall that this is the probability distribution for x) is obtained by

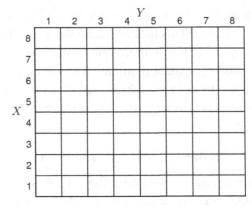

Figure 10.1. A grid corresponding to the possible outcomes (samples) for two discrete random variables X and Y. Each square in the grid corresponds to a sample in the sample space Ω.

using the inverse of f to map the set of values $(-\infty, x]$ to the corresponding event in the sample space, and then using $\mathbb{P}$ to get the probability. We can write this as

$$D(x) \equiv \text{Prob}(X \in (\infty, x]) = \mathbb{P}\left[f^{-1}\{(\infty, x]\}\right]. \qquad (10.8)$$

Since $D(x)$ is the probability distribution for X, the probability density is

$$P(x) = \frac{d}{dx}\mathbb{P}\left[f^{-1}\{(\infty, x]\}\right]. \qquad (10.9)$$

Independent variables and σ-algebras

In modern probability theory one thinks about the independence of two random variables in terms of σ-algebras. This is much simpler than it sounds. Consider two discrete random variables X and Y, each of which take the values one through ten. The total set of possibilities is the one hundred possible combinations of their values, (x, y). Since each sample corresponds to one of these combinations, there are one hundred samples in the sample space Ω.

Figure 10.1 displays a grid in which each square represents one of the samples. Now consider the variable X alone. The event $X = 1$ corresponds to any one of the samples in the top row. The probability that $X = 1$ therefore corresponds to the sum of the probabilities of the samples in the top row. Another way to say this is: the probability that $X = 1$ is the value given by the probability measure for the set of samples consisting of the entire top row of samples. Correspondingly, the probability that $X = n$ is given by the probability measure for the set consisting of the nth row. This means that to calculate the probability for any value of X, or subset of the values of X, we only need consider the sets of samples that consist of *full rows*, or the unions of full rows. The collection of all these "row sets" forms a σ-algebra. This σ-algebra is not the full σ-algebra for the sample space, since there are many sets of samples (the columns, for example) that it does not include. It is therefore a subalgebra of the sample space. This subalgebra allows us to calculate all the probabilities relating to the random variable X alone, and is thus sufficient to "support" X. This subalgebra is referred to as the σ-algebra *generated* by X. If σ is a σ-algebra, the notation $\sigma(X)$ means that σ is the algebra generated by the random variable X.

We have now seen that each independent random variable is described by a subalgebra of the σ-algebra of the sample space. In the example in Figure 10.1, X is described by the subalgebra of the rows, and Y by the subalgebra of the columns. In modern probability theory, one defines the independence of two (or more) random variables in terms of the independence of their respective σ-algebras. While this is not so important for our purposes, two σ-algebras, σ_X and σ_Y, are independent if the probability measure of the intersection of any subset in σ_X with any subset in σ_Y, is the *product* of the measures of the two subsets. In symbols this is

$$\mathbb{P}(A \cap B) = \mathbb{P}(A)\mathbb{P}(A), \quad \text{for all} \quad A \in \sigma_X \quad \text{and} \quad B \in \sigma_Y. \quad (10.10)$$

This is another way of saying that the joint probability density of X and Y is the product of their respective probability densities.

Note that the σ-algebra that describes both X and Y subsumes the separate σ-algebras for each. By this we mean that all the subsets that are contained in the σ-algebra for X are also contained in the σ-algebra that describes the possible values of X and Y. In our example above this is the σ-algebra of the whole sample space, Ω. However, if the sample space contained three independent random variables, then the σ-algebra for X and Y would merely be a subalgebra of the full σ-algebra for Ω.

Conditional expectation

In classical probability theory, one usually thinks about the relationship between two random variables in terms of their joint probability density, or conditional probability densities. Mathematicians often think instead in terms of the *expectation* value of one variable given the value of another. The expectation value of X given a specific value for Y is called the *conditional expectation value* of X, conditioned on the value of Y, and is denoted by $E[X|Y]$. The conditional expectation value is actually itself a random variable. This is because $E[X|Y]$ has a specific value for each value of Y. Thus $E[X|Y]$ is a random variable that has the *same* σ-algebra as Y.

Technical note on random variables. Say that we have a random variable x given by the function f. Since f maps each sample point to a value, it also maps a specific *set* of sample points to a specific *set* of values. If we want to calculate the probability of that x falls within a certain set of values, then f (strictly, the inverse of f) maps this set back to an event in the sample space, and the measure tells us the probability of that event. So the procedure of mapping back to the sample space, and using the measure defined on this space (technically, on the σ-algebra of this space), actually provides a measure on the space of values of x. We would like this measure to assign a value to every Borel set in this space. This means that the inverse of f must map all the Borel sets to the σ-algebra (or to a σ-subalgebra) of the sample space, $\mathcal{F}$. A function that does this is called a ***measurable function***. Thus, technically, a random variable is defined by a measurable function on the sample space. If a random variable is described by the σ-algebra Σ, then the function that defines it is referred to as being Σ-*measurable*.

10.7 Stochastic processes: *filtrations* and *adapted processes*

We now have almost all the jargon we need to understand research articles on stochastic processes employing modern probability theory. The final piece is the terminology related to stochastic processes themselves. Consider now a discretized stochastic process, $X(t)$, whose increment in each time interval Δt is ΔX_n. As usual we have N time intervals, and we label the increments $\Delta X_0, \ldots, \Delta X_{N-1}$. We will also denote the value of X at time $n\Delta t$ by $X_n = X(n\Delta t) = \sum_0^{n-1} \Delta X_j$. At time $t = 0$ we do not know the value of any of the increments of X; at that time they are all still in future. As time passes, all of the increments take on specific values, one after the other. At time $t = n\Delta t$ we know the values of the ΔX_n for $n = 0, \ldots, n - 1$.

The sample space for this stochastic process includes all the possibilities for all of the random variables ΔX_n. We will denote the σ-algebra for each of these variables by $\sigma(\Delta X_n)$, and the combined σ-algebra for the set of variables $\Delta X_0, \ldots, \Delta X_n$ as $\mathcal{F}_n$. Note that the σ-algebra $\mathcal{F}_n$ contains all the events that can happen up until time $(n + 1)\Delta t$. That is, if we want to know the probability for anything that happens up until time $(n + 1)\Delta t$, we only need the σ-algebra $\mathcal{F}_n$. Recall from Section 10.6 that the σ-algebra for a set of random variables contains the σ-algebras for each of them. To signify that one σ-algebra, Σ_1, is contained within another, Σ_2, we use the symbol $\subseteq$, and write $\Sigma_1 \subseteq \Sigma_2$. We therefore have

$$\mathcal{F}_0 \subseteq \mathcal{F}_1 \subseteq \cdots \subseteq \mathcal{F}_{n-1} \subseteq \cdots . \qquad (10.11)$$

A sequence of σ-algebras that describes what information we have access to at successive times is called a ***filtration***. The sequence of σ-algebras $\mathcal{F}_n$ is thus a filtration. Stochastic processes are therefore described by a sample space Ω, an all-encompassing σ-algebra $\mathcal{F}$, a probability measure $\mathbb{P}$, *and* a filtration, which is usually denoted by $\{\mathcal{F}_n\}$ (or just $\mathcal{F}_n$). This collection of objects is called a *filtered probability space*, or just *filtered space*, and usually written as $(\Omega, \mathcal{F}, \mathcal{F}_n, \mathbb{P})$

For a continuous stochastic process, $X(t)$, we merely replace the discrete index n with the time t. So $\mathcal{F}_t$ is the σ-algebra for all the infinitesimal increments of X up until time t. We also have $\mathcal{F}'_t \subseteq \mathcal{F}_t$ when $t' \leq t$, and the filtered space is now written as $(\Omega, \mathcal{F}, \mathcal{F}_t, \mathbb{P})$. An **adapted process**, $Y(t)$, (also referred to as being *adapted to* $\mathcal{F}_t$) is a process that at time t is only a function of the increments of $X(t)$ up until that time. Thus an adapted process at time t is fully described by $\mathcal{F}_t$.

To summarize: a *filtration* is a sequence of σ-algebras, one for each time t, and the σ-algebra at time t describes all the stochastic increments up until that time. An adapted process, $X(t)$, is one that is a function only of information available up until time t.

10.7.1 Martingales

The term "martingale" often arises in mathematics papers, and is a very simple concept. (Incidentally, this term derives from a French acronym for a gambling strategy.) A **martingale** is a stochastic process whose expectation value for the next time-step is the same as its value at the present time. So if the value of the process $X(t)$ at time t is *known*, then X is a martingale if

$$\langle X(t + dt) \rangle = X(t). \tag{10.12}$$

The above equation looks a bit odd (strictly it is an "abuse of notation"), because we do not usually think of the value of a stochastic process at time t as being known: we usually think of $X(t)$ as representing a random variable at time t, rather than the *value* of this variable. However, once we have *arrived* at time t, then this value is known. A more explicit way to write Eq. (10.12) is

$$E[X(t + dt)|X(t)] = E[X(t)|X(t)], \tag{10.13}$$

since the expectation value of $X(t)$ *given* $X(t)$ *is* the value of X at time t.

There are also sub-martingales, super-martingales and semi-martingales. A process $X(t)$ is a **sub-martingale** if

$$E[X(t + dt)|X(t)] \geq E[X(t)|X(t)], \tag{10.14}$$

and is a **super-martingale** if

$$E[X(t + dt)|X(t)] \leq E[X(t)|X(t)]. \tag{10.15}$$

This terminology comes from a relationship with super- and sub-harmonic functions. The definition of semi-martingales is more technical, and involves the continuity of stochastic integrals of the process in question [33]. We omit this definition here, but note that all Levy processes are semi-martingales.

10.8 Translating the modern language

With the above jargon in hand, the reader should be able to read many research articles written using the modern formalism. To make this easier, we now give as examples a couple of excerpts taken from such works. Since we have drawn these from control theory, the following simple background knowledge will be useful: the evolution of a noisy dynamical system can be controlled by continually changing the forces on the system. This amounts to changing one or more parameters that appear in the differential equations for the system. If we denote the set of parameters by the vector $\mathbf{c}$, and the state of the system by the vector $\mathbf{x}$, then the equations for the system might be

$$dx = f(\mathbf{x}, \mathbf{c}, t)dt + g(\mathbf{x}, \mathbf{c}, t)dW. \tag{10.16}$$

To control the system we make $\mathbf{c}$ a function of time. More specifically, we make $\mathbf{c}(t)$ a function of $\mathbf{x}(t)$. This control procedure is referred to as *feedback control*, because we are "feeding back" our knowledge of where the system is at time t to modify the dynamics, with the hope of correcting for the effects of noise.

Now to our first example. The following introductory passage is taken from the text on control theory by Alain Bensoussan [39] (page 23).

Let $(\Omega, \mathcal{A}, P)$ be a probability space on which is given
$w(\cdot)$ a Wiener process with values in R^n, with covariance matrix $Q(\cdot)$.
Let $\mathcal{F}^t = \sigma(w(s), s \leq t)$. The process $w(\cdot)$ will be the unique source of noise in the model, and we assume full information, i.e. $\mathcal{F}^t$ represents the set of observable events at time t.
An admissible control is a process $v(\cdot)$ adapted to $\mathcal{F}^t$, which is square integrable.
Let $v(\cdot)$ be an admissible control; the corresponding state is the solution of

$$dx = (F(t)x + G(t)v + f(t))dt + dw, \tag{10.17}$$

$$x(0) = x_0. \tag{10.18}$$

Now that we know the concepts and associated jargon of modern probability theory, this introductory passage is not hard to follow. The first line states that $(\Omega, \mathcal{A}, P)$ is a probability space, so we infer that the author is using $\mathcal{A}$ to denote the σ-algebra of events, and P instead of $\mathbb{P}$ for the probability measure. The next line tells us that "$w(\cdot)$ is a Wiener process". The statement that w takes values in R^n means that it is actually a *vector* of n Wiener processes, and that the covariance matrix that gives the correlations between these Wiener processes is denoted by C. The "dot" given as the argument to w just means that w is a function of

something – since it is a vector of Wiener processes we can assume that this
something is time, t. The covariance matrix is thus also a function of t. The next
line tells us that $\mathcal{F}^t = \sigma(w(s), s \le t)$. Since we can already guess that $\{\mathcal{F}^t\}$ is a
filtration that supports a stochastic process, we don't need to pay any attention
to the expression "$\sigma(w(s), s \le t)$". Nevertheless, this expression states that $\mathcal{F}^t$
is the σ-algebra generated by the vector Wiener process w up until time t. This
means, as we expect, that $\mathcal{F}^t$ is the tth element of the sequence of σ-algebras (the
filtration) that supports w. The text then says that "we assume full information".
What this means can be inferred from the phrase that follows: "$\mathcal{F}^t$ represents the
set of observable events at time t". This means that at time t the observer, who is
also the controller, knows the realizations of all the Wiener processes up until t.
Finally, the text states that an "admissible control" is a process adapted to the
filtration $\mathcal{F}^t$. We now know that this just means that $v(t)$ depends only on the
realizations of the Wiener processes up until time t. The physical meaning of $v(t)$
(that is, its role in the control process) can then be inferred from the final sentence.
This says that the "state" (meaning the state of the system to be controlled) is the
solution to Eq. (10.17). Examining this equation we find that $v(t)$ appears in the
equation that determines the motion of the system, and thus represents a set of
"control parameters" that can be modified as time goes by.

We now turn to a more challenging example, in which the mathematical style
of presentation makes things seem even more obscure. The best way to read the
introduction to such papers, I find, is to jump around, picking up the necessary bits
of information from here and there to piece the whole together. The following is a
(slightly modified) version of an excerpt taken from an article on optimal control
theory by Zhou, Yong, and Li [40].

We consider in this paper stochastic optimal control problems of the following kind. For
a given $s \in [0, 1]$, by the set of admissible controls $U_{ad}[s, 1]$ we mean the collection of
(i) standard probability spaces $(\Omega, \mathcal{F}, P)$ along with m-dimensional Brownian motions
$B = \{B(t) : s \le t \le 1\}$ with $B(s) = 0$ and (ii) Γ-valued $\mathcal{F}_t^s$-adapted measurable processes
$u(\cdot) = \{u(t) : s \le t \le 1\}$, where $\mathcal{F}_t^s = \sigma\{B(r) : s \le r \le t\}$ and Γ is a given closed set in
some Euclidean space R^n. When no ambiguity arises, we will use the shorthand $u(\cdot) \in$
$U_{ad}[s, 1]$ for $(\Omega, \mathcal{F}, P, B, u(\cdot)) \in U_{ad}[s, 1]$. Let $(s, y) \in [0, 1) \times R^d$ be given. The process
$x(\cdot) = \{x(t) : s \le t \le 1\}$ is the solution of the following Ito stochastic differential equation
(SDE) on the filtered space $(\Omega, \mathcal{F}, \mathcal{F}_t^s, P)$:

$$\begin{cases} dx(t) = f[t, x(t), u(t)]dt + \sigma[t, x(t), u(t)]dB(t), \\ x(s) = y. \end{cases} \qquad (10.19)$$

The process $x(\cdot)$ is called the response of the control $u(\cdot) \in U_{ad}[s, 1]$, and $(x(\cdot), u(\cdot))$ is
called an admissible pair.

When we begin reading, the first definition we encounter is "the set of admissible
controls". From the phrase "by the set of admissible controls $U_{ad}[s, 1]$ we mean the

collection of (i) standard probability spaces $(\Omega, \mathcal{F}, P)$ along with . . .", it appears that this might be the whole collection of all possible probability spaces. This doesn't appear to make much sense, so let's read on. Further down the text, we find the fragment "for $(\Omega, \mathcal{F}, P, B, u(\cdot)) \in U_{ad}[s, 1]$", and this tells us that an element of the "set of admissible controls" is the collection of five objects given by $(\Omega, \mathcal{F}, P, B, u(\cdot))$. By going back up the text, we see that $(\Omega, \mathcal{F}, P)$ is a probability space, and we know what that is. The author is therefore using P rather than $\mathbb{P}$ to denote the probability measure. So a single admissible control is a probability space, with two other things, B and $u(\cdot)$. Another fragment tells us that B is an "m-dimensional Brownian motion". A "Brownian motion" is an alternative name for a Wiener process, and "m-dimensional" means that it is a vector of Wiener processes (which we will call a "vector Wiener process" for short). It also states that $u(\cdot)$ is an "$\mathcal{F}_t^s$-adapted measurable process". (As above, the "dot" inside the parentheses just means that u is a function of something.) We know that u is a "process", and so it is presumably a function of time. We know also that u is "adapted to $\mathcal{F}_t^s$", so this means that u is a stochastic process, and that $\mathcal{F}_t^s$ is a filtration. It is reasonable to conclude that the filtration supports the process B, and thus that $u(t)$ is some function of this Wiener process. (We also wonder why the filtration has a superscript as well as a subscript. Looking back we glean from the segment "$\mathcal{F}_t^s = \sigma\{B(r) : s \leq r \leq t\}$" that $\mathcal{F}_t^s$ is the σ-algebra that supports the process B between times s and t. So the superscript denotes the initial time.)

We now see that an "admissible control" is a vector Wiener process B, and some function of B called u, along with the probability space that supports them. But the physical meaning of u is not yet clear. We now read on, and find that a new process, $x(t)$ is defined. We are told that this is the solution to an "Ito stochastic differential equation (SDE) on the filtered space $(\Omega, \mathcal{F}, \mathcal{F}_t^s, P)$". We know that the "filtered" space is just the probability space that supports our vector of Wiener processes. We now look at the stochastic equation for x, and see that it is driven by the increment of the vector Wiener process, dB. We also see that $u(t)$ is a parameter in the equation. Now it all starts to make sense – $u(t)$ is something that affects the motion of x, and can therefore be used to control x. It makes sense then that $u(t)$ is some function of the Wiener process up until time t, since the control we choose at time t will depend on the value of x, and $x(t)$ depends of the Wiener processes up until time t. Reading on, this is confirmed by the phrase below Eq. (10.19), "process $x(\cdot)$ is called the response of the control $u(\cdot) \in U_{ad}[s, 1]$". So it is really only $u(t)$ that is the "admissible control", not the entire collection $(\Omega, \mathcal{F}, P, B, u(\cdot))$. The authors of Ref. [40] clearly feel the need to include the probability space along with $B(t)$ and $u(t)$ as part of the definition of the "admissible control", presumably because $u(t)$ depends on $B(t)$, and $B(t)$ is supported by the rest. We also see from Eq. (10.19) that the value of x is given at time s. Returning the definition of u, we see that $u(t)$ is defined between the times s and unity. So the control problem is to control x in the time-interval $[s, 1]$, with the initial condition $x(s) = y$. The text also states that u is "Γ-valued". To find out what this means we locate the definition of Γ, being "a closed set in R^n". This means that u is an n-dimensional vector that comes from a "closed set". Thus the values of the elements of u have some arbitrary (but finite) bounds placed upon them. (Note: if we were to read further in the article we would find that the function $x(t)$ is a vector, but we will stop here.)

We can now rewrite the above excerpt in plain language.

Here we consider the problem of controlling a dynamical system described by the stochastic equation

$$\mathbf{dx}(t) = \mathbf{f}[t, \mathbf{x}(t), \mathbf{u}(t)]dt + \sigma[t, \mathbf{x}(t), \mathbf{u}(t)]\mathbf{dB}. \tag{10.20}$$

Here $\mathbf{x}(t)$ is the state of the system, which will in general be a vector, $\mathbf{dB} = (dB_1, \ldots, dB_n)$ is a vector of independent Wiener processes, all of which satisfy the Ito calculus relation $dB_j^2 = dt$, and the functions $\mathbf{f}$ and σ are arbitrary. The vector $\mathbf{u}$ is the set of parameters that we can modify to control the system. Thus at time t, the control parameters $\mathbf{u}(t)$ can in general be a function of all the Wiener increments up until t. The control problem is trivial unless the control parameters are bounded. We allow these bounds to be arbitrary, so $\mathbf{u}(t)$ merely lies in some closed set Γ in R^n. We will consider controlling the system from an initial time $t = s \geq 0$ to the fixed final time $t = 1$. Naturally the stochastic equation is supported by some filtered probability space $(\Omega, \mathcal{F}, \mathcal{F}_t, \mathbb{P})$.

In the above passage I say that "the functions $\mathbf{f}$ and σ are arbitrary". Mathematicians would complain that this is imprecise – what I really mean is that $\mathbf{f}$ and σ are any functions that are, say, twice differentiable. The primary difference between the language of natural scientists and mathematicians is that natural scientists take it as automatically implied in their papers that, for example, any functions that appear are sufficiently differentiable for the purposes for which they are used.

My own view is that the language favored by mathematicians, while precise, usually adds little that is of value to natural scientists. In this I am not alone; the great Edwin T. Jaynes argues this point of view eloquently in his classic work *Probability Theory: The Logic of Science* [1]. This exposition appears under the headings "What am I supposed to publish?", and "Mathematical courtesy", on pages 674–676. Jaynes goes further and suggests that modern writers on probability theory could shorten their works considerably by including the following proclamation (with perhaps the addition of another sentence using the terms "sigma-algebra, Borel field, Radon–Nikodým derivative") on the first page.

Every variable x that we introduce is understood to have some set X of possible values. Every function $f(x)$ that we introduce is understood to be sufficiently well-behaved so that what we do with it makes sense. We undertake to make every proof general enough to cover the applications we make of it. It is an assigned homework problem for the reader who is interested in the question to find the most general conditions under which the result would hold.

Further reading

For a rigorous introduction to modern probability theory, we recommend the concise book by David Williams, *Probability with Martingales* [35]. This introduces the concepts of modern probability theory as well as stochastic processes. It does not explicitly consider continuous-time processes, but since all the terminology and concepts transfer directly to continuous-time, little is lost. Two texts that do specifically include continuous-time Wiener noise, use modern probability theory, and focus on modeling and applications rather than mathematical detail, are *Financial Modelling with Jump Processes* by Cont and Tankov [33] and *Stochastic Differential Equations: An Introduction with Applications* by Bernt Øksendal [11]. The first of these also contains an introduction to the concepts of modern probability theory. Texts that are more focussed on the mathematics of stochastic processes are *Brownian Motion and Stochastic Calculus* by Karatzas and Shreve [12], and the two-volume set *Diffusions, Markov Processes, and Martingales* by Rogers and Williams [13].

Appendix A

Calculating Gaussian integrals

To calculate means and variances for Gaussian random variables, and the expectation values of exponential and Gaussian functions of Gaussian random variables, we need to do integrals of the following form

$$\int_{-\infty}^{\infty} x^n e^{-\alpha x^2 + \beta x} dx. \tag{A1}$$

First let us solve this integral when $n = 0$. To do this we need to complete the square in the exponential, which means writing

$$-\alpha x^2 + \beta x = -\alpha \left(x^2 - \frac{\beta}{\alpha} x \right). \tag{A2}$$

Next we write the expression in the brackets as $(x + a)^2 - b$. That is, we solve

$$(x - a)^2 + b = x^2 - \frac{\beta}{\alpha} x \tag{A3}$$

for a and b. This gives

$$a = \frac{\beta}{2\alpha} \quad \text{and} \quad b = -\frac{\beta^2}{4\alpha^2}. \tag{A4}$$

Putting this back into Eq. (A2) we have

$$-\alpha x^2 + \beta x = -\alpha \left[\left(x - \frac{\beta}{2\alpha} \right)^2 - \frac{\beta^2}{4\alpha^2} \right]$$

$$= -\alpha \left(x - \frac{\beta}{2\alpha} \right)^2 + \frac{\beta^2}{4\alpha}. \tag{A5}$$

The integral becomes

$$\int_{-\infty}^{\infty} e^{-\alpha x^2 + \beta x} dx = \int_{-\infty}^{\infty} e^{-\alpha [x - \beta/(2\alpha)]^2 + \beta^2/(4\alpha)} dx$$

$$= e^{\beta^2/(4\alpha)} \int_{-\infty}^{\infty} e^{-\alpha [x - \beta/(2\alpha)]^2} dx. \tag{A6}$$

181

We now perform the substitution $v = x - \beta/(2\alpha)$, and the integral becomes

$$e^{\beta^2/(4\alpha)} \int_{-\infty}^{\infty} e^{-\alpha[x-\beta/(2\alpha)]^2} dx = e^{\beta^2/(4\alpha)} \int_{-\infty}^{\infty} e^{-\alpha v^2} dv. \tag{A7}$$

We now use the formula [37],

$$\int_{-\infty}^{\infty} e^{-\alpha v^2} dv = \sqrt{\frac{\pi}{\alpha}}, \tag{A8}$$

and so the solution to the integral is

$$\int_{-\infty}^{\infty} e^{-\alpha x^2 + \beta x} dx = \sqrt{\frac{\pi}{\alpha}} e^{\beta^2/(4\alpha)}. \tag{A9}$$

To do the integral for $n = 1$, we first complete the square and make the substitution as above. This time we get

$$\int_{-\infty}^{\infty} x e^{-\alpha x^2 + \beta x} dx = e^{\beta^2/(4\alpha)} \int_{-\infty}^{\infty} [v + \beta/(2\alpha)] e^{-\alpha v^2} dv$$

$$= e^{\beta^2/(4\alpha)} \left(\int_{-\infty}^{\infty} v e^{-\alpha v^2} dv + \frac{\beta}{2\alpha} \int_{-\infty}^{\infty} e^{-\alpha v^2} dv \right). \tag{A10}$$

The first integral is zero because the integrand is anti-symmetric, and so the answer is

$$\int_{-\infty}^{\infty} x e^{-\alpha x^2 + \beta x} dx = \sqrt{\frac{\pi}{\alpha}} \frac{\beta}{2\alpha} e^{\beta^2/(4\alpha)}. \tag{A11}$$

Finally, to do $n = 2$ we again complete the square and perform the substitution as before. This time we get integrals of the form we have already encountered above, plus the integral

$$\int_{-\infty}^{\infty} v^2 e^{-\alpha v^2} dv. \tag{A12}$$

To do this we integrate by parts by splitting the integrand into $f(v) = v$ and $g = v e^{-v^2}$. We can do this because we can integrate $g(v)$ using the substitution $u = v^2$. The result is

$$\int_{-\infty}^{\infty} v^2 e^{-\alpha v^2} dv = \frac{-v e^{-\alpha v^2}}{2\alpha} \bigg|_{-\infty}^{\infty} + \frac{1}{2\alpha} \int_{-\infty}^{\infty} e^{-\alpha v^2} dv. \tag{A13}$$

The first term is zero, and thus

$$\int_{-\infty}^{\infty} v^2 e^{-\alpha v^2} dv = \sqrt{\frac{\pi}{\alpha}} \frac{1}{2\alpha}. \tag{A14}$$

To do the integral in Eq. (A1) for $n > 2$, one completes the square and makes the substitution $v = x - \beta/(2\alpha)$ as above. Then one has integrals of the form

$$\int_{-\infty}^{\infty} v^n e^{-\alpha v^2} dv. \tag{A15}$$

To do these one integrates by parts, splitting the integrand into $f(v) = v^{n-1}$ and $g(v) = v e^{-v^2}$. One does this repeatedly until all integrals are reduced to either $n = 1$ or $n = 0$, for which we know the answer.

Finally, we note that the above techniques also work when the coefficients of the quadratic in the exponential are complex, because

$$\int_{-\infty+iq}^{\infty+iq} e^{-\alpha v^2} dv = \sqrt{\frac{\pi}{\alpha}} \tag{A16}$$

for any real q and complex α with $\text{Re}[\alpha] > 0$ [38].

The following general formulae are also useful:

$$\int_0^\infty x^{2n} e^{-\alpha x^2} dx = \frac{\sqrt{\pi}}{(2\sqrt{\alpha})^{2n+1}} \frac{(2n)!}{n!}, \tag{A17}$$

$$\int_0^\infty x^{2n+1} e^{-\alpha x^2} dx = \frac{n!}{2\alpha^{n+1}}. \tag{A18}$$

References

[1] E. T. Jaynes, *Probability Theory: The Logic of Science* (Cambridge University Press, Cambridge, 2003).

[2] E. T. Jaynes, in *E. T. Jaynes: Papers on Probability, Statistics, and Statistical Physics*, edited by R. D. Rosenkrantz (Dordrecht, Holland, 1983).

[3] S. M. Tan, *Linear Systems*, available at http://home.comcast.net/~szemengtan/

[4] A. V. Oppenheim and A. S. Willsky, *Signals and Systems* (Prentice Hall, 1996).

[5] J. I. Richards and H. K. Youn, *The Theory of Distributions: A Nontechnical Introduction* (Cambridge University Press, Cambridge, 1995).

[6] C. E. Shannon and W. Weaver, *The Mathematical Theory of Communication* (University of Illinois Press, Illinois, 1998).

[7] C. H. Edwards and D. E. Penney, *Differential Equations and Linear Algebra* (Prentice Hall, 2008).

[8] D. C. Lay, *Linear Algebra and Its Applications* (Addison Wesley, Tappan, NJ, 2005).

[9] M. Newman and R. C. Thompson, *Math. Comput.* **48**, 265 (1987).

[10] J. A. Oteo, *J. Math. Phys.* **32**, 419 (1991).

[11] B. Øksendal, *Stochastic Differential Equations: An Introduction with Applications* (Springer, New York, 2007).

[12] I. Karatzas and S. E. Shreve, *Brownian Motion and Stochastic Calculus* (Springer, New York, 2008).

[13] L. C. G. Rogers and D. Williams, *Diffusions, Markov Processes, and Martingales*, vols. 1 and 2 (Cambridge University Press, Cambridge, 2000).

[14] K. Jacobs and P. L. Knight, *Phys. Rev.* A **57**, 2301 (1998).

[15] R. M. Howard, *Principles of Random Signal Analysis and Low Noise Design: The Power Spectral Density and its Applications* (Wiley-IEEE Press, New York, 2002).

[16] R. N. McDonough and A. D. Whalen, *Detection of Signals in Noise* (Academic Press, New York, 1995).

[17] S. M. Tan, *Statistical Mechanics*, available at http://home.comcast.net/~szemengtan/

[18] B. Lukić, S. Jeney, C. Tischer, A. J. Kulik, L. Forró, and E.-L. Florin, *Phys. Rev. Lett.* **95**, 160601 (2005).

[19] Z. Schuss, A. Singer, and D. Holcman, *Proc. Nat. Acad. Sci. U.S.A.* **104**, 16098 (1995).

[20] P. Wilmott, S. Howison, and J. Dewynne, *The Mathematics of Financial Derivatives: A Student Introduction* (Cambridge University Press, Cambridge, 1995).

[21] P. E. Kloeden and E. Platen, *Numerical Solution of Stochastic Differential Equations* (Springer, New York, 2000).

[22] W. H. Press, S. A. Teukolsky, W. T. Vetterling, and B. P. Flannery, *Numerical Recipes in C: The Art of Scientific Computing* (Cambridge University Press, Cambridge, 1992).

[23] C. W. Gardiner, *Handbook of Stochastic Methods*, second edition (Springer, New York, 1990).

[24] A. Hagberg and E. Meron, *Nonlinearity* **7**, 805 (1994a).

[25] A. Hagberg and E. Meron, *Chaos* **4**, 477 (1994b).

[26] A. Hagberg, Ph.D. dissertation, University of Arizona, Tucson (1994).

[27] R. E. Goldstein, D. J. Muraki, and D. M. Petrich, *Phys. Rev.* E **53**, 3933 (1996).

[28] C. Chuan-Chong and K. Khee-Meng, *Principles and Techniques in Combinatorics* (World Scientific, 1992).

[29] E. P. Odum, *Fundamentals of Ecology* (Saunders, Philadelphia, 1953).

[30] L. Gammaitoni, P. Hanggi, P. Jung, and F. Marchesoni, Stochastic resonance, *Rev. Mod. Phys.* **70**, 223 (1998).

[31] M. Kanter, *Ann. Prob.* **3**, 697 (1975).

[32] J. Chambers, C. Mallows, and B. Stuck, *J. Am. Stat. Assoc.* **71**, 340 (1976).

[33] R. Cont and P. Tankov, *Financial Modelling with Jump Processes* (Chapman & Hall, London, 2004).

[34] R. Metzler and J. Klafter, *Phys. Rep.* **339**, 1 (2000).

[35] D. Williams, *Probability with Martingales* (Cambridge University Press, Cambridge, 1991).

[36] M. Clark, *Paradoxes from A to Z* (Routledge, 2007).

[37] A. E. Taylor and W. R. Mann, *Advanced Calculus,* 3rd edn, pp. 680–681 (Wiley, 1983).

[38] I. S. Gradshteyn, I. M. Ryzhik, A. Jeffrey, and D. Zwillinger, *Table of Integrals, Series, and Products* (Academic Press, 2000).

[39] A. Bensoussan, *Stochastic Control of Partially Observable Systems* (Cambridge University Press, Cambridge, 2004).

[40] X. Y. Zhou, J. Yong, and X. Li, Stochastic verification theorems within the framework of Viscosity Solutions, *SIAM J. Cont. Optim.* **35**, 243–253 (1997).

Index